DICTIONARY OF
CROP SCIENCE

DICTIONARY OF CROP SCIENCE

by

Dr. R.P. Chandola

2010
DAYA PUBLISHING HOUSE
Delhi - 110 035

ISBN 81-7035-656-3
ISBN 978-81-7035-656-1

Published by : **Daya Publishing House**
1123/74, Deva Ram Park
Tri Nagar, Delhi - 110 035
Phone: 27383999
Fax: (011) 23260116
e-mail : dayabooks@vsnl.com
website : www.dayabooks.com

Showroom : 4760-61/23, Ansari Road, Darya Ganj,
New Delhi - 110 002
Phone: 23245578, 23244987

Laser Typesetting : **Classic Computer Services**
Delhi - 110 035

Printed at : **Chawla Offset Printers**
Delhi - 110 052

PRINTED IN INDIA

Preface

Words in any language have always taken a major role in the advancement of human civilization. With the passing of time many words automatically added and enriched the various cultures towards provisions of a better, secured and longer life of all the living beings on this planet–the earth. Words have been used in the area of education which in turn has played a significant role towards improvement and upgradation of human life. In this regard scientific knowledge acquired through modern scientific research and education has influenced the human life tremendously vis-à-vis all other connected human activities.

Agricultural science including animal science as a particular way of life as adopted by human beings and also moved in the same direction as other sciences and also acts made good progress only primarily by means up many words with passing out of time. As far as scientific development for further improvement is concerned, it firstly evolved the scheme of making dictionaries to make it possible for easy comprehension of human behaviour, actions and reactions towards a civilized society. Such works also helped and emerged as one of the most important way in achieving an easy consultation for an achievement of rapid and significant knowledge of the different disciplines of agricultural science. In the present vastness of ocean of knowledge of any subject, it is difficult and cumbersome find the correct words for the different functions of agricultural science. The situation has become more tedious as the worlds pertain to the scientific and technical notations of various agricultural branches of knowledge have become numerous.

With the allure in view, six volumes relevant to the various subjects of agriculture *e.g.* Animal Science, Entomology (insect pest study), Crop Science (growing various agricultural field crops, agronomy), Plant Science (plant breeding and genetics), Plant Pathology (plant diseases), and Soil Science have been prepared with all scientific names (terms) used in agricultural education and research with comprehensive notes for various terms. Whenever possible some entries have been illustrated to make an easy understanding possible. Such a dictionary should also be used as a quick and easy reference to students and other concerned groups.

Dr. R.P. Chandola

A

Absolute humidity The amount of water vapour per unit volume of air, usually expressed in grams per cubic metre, Compare relative humidity, specific humidity.

Absolute temperature (thermodynamic temperature) A temperature measured on a scale based on the lowest temperature that is physically possible (~273.15°C) known as absolute zero or 0 kelvin (K). It was formerly measured in degrees absolute (°A) but is now measured in kelvins (the degree sign is now officially omitted). For meteorological purposes, the absolute temperature is taken as the Celsius temperature plus 273°C, for example – 10°C is 263 K and 100°C (boiling point is 273 K).

Acacia Arabica This plant can be used successfully as a hedge for fields and orchards because of its thorny nature.

Acanthospermum hispidum Fam. Compositeae. Any method may be used which is effective in preventing flowering to control this weed.

A. carambola Kamarak, Fam. Oxalideceae. Agood raw material for making jelly.

Accumulated temperature A measure of the variation of temperature from a given norm for a particular place over a specified period. Measured in degree/days. This is most commonly used in determining whether an area is suitable for the cultivation of particular crop. Over a given period (*e.g.* 30 days), the difference between the base value is taken as the temperature below which plant growth is inhibited, are added together, the result indicating the effectiveness of temperature over the given period.

Accumulation of toxic effects of herbicides If the instructions given on the products are faithfully followed, there is no possibility of the pesticides or herbicides leaving toxic residues in food products.

Achrus sapota Also known as sapodilla, sapota or chikoo, Family sapotaceae. A tropical fruit.

Acclimatize To adapt or become accustomed to a new environmental conditions. Usually it means to adapt to a different climatic regime though sometimes the plants adapt to change soil conditions.

Achyranthes aspera Fam. Amaranthaceae. A serious weed of summer season crops. Can be controlled by handpicking.

Acidity and basicity of fertilizers The acidifying capacity in terms of pounds of $CaCO_3$ for every 1 per cent of N in 1 ton of fertilizer when applied to the soil is approximately:

Ammonium sulphate	107	High grade tankage	15
Ammo-phos	107	Fish scrap	31
Monocalcium phosphate	107	Cotton seed cake	29
Urea	36	Castor pomace	18

The basicity in the same terms for certain fertilizers is:

Nitrate of soda	36	Tobacco stems	86
Calcium Cyanamid	37	Cocoa meal	12

Acid tolerance in crops Many of the principal crops and vegetables are sensitive to acid soils and suffer injury when grown on them.

Acre-furrow-slice The extent of soil in one acre and 1 inch deep. These are used for computation for certain items as organic matter, actual mineral constituents and the volume weight which is necessary in the calculation of the total pore space present in soil. Acre furrow-slice contains 2,000,000 lbs, of soil material.

Acre-inch Represents the quantity of water that will cover an acre of surface soil one inch deep. One acre-inch of water weights about 101 tons. Twelve acre-inches are equivalent to one acre-foot. A cusec of water flowing for one hour supplies approximately one acre-inch.

Activated sludge The effluent from the settling tank of sewage carries a large amount of objectionable colloidal matter, and the deposit of sludge that settles in the tank is of small manurial value and is often of offensive character. These defects can be removed by thoroughly aerating the sewage in the settling tank by blowing air through it. The sludge that settles at the bottom in this process is called 'activated sludge.' It has the remarkable property of helping the rapid oxidation of the organic matter in fresh sewage.

Adiabatic process An atmospheric process in which changes take place in the pressure and volume of a body of gas without any actual loss or gain of heat from outside.

Adaptability Capacity of an animal or plant to grow well in a given environment.

Adiabatic rate lapse If the rising air is saturated, it cools at the saturated adiabatic lapse rate. This differs from the dry adiabatic rate because of latent heat released by condensation of vapour.

Adjusted error sum of square (Sta) For this, we noticed to adjust a lot yields (y) for the influence of the plant number (x) in so far as its random variation from plot to plot is concerned. It is therefore the relation as reflected in the covariance between y and x, the error line which must be employed for the adjustment. The relation, in so far as it is linear, is expressed by the linear regression coefficient of x on y. This has a value equal to Sp(xy)/D.F-error and carries a single degree of freedom with a corresponding sum of squares given by

$$\frac{[S.P.(xy)]^2}{S.S.(x^2)}$$

By removing this quantity from the error, sum of squares for y we obtain the sum of squares for residual random variation in y called the adjusted error sum of squares.

Adoretus lesiopygus* and *A. versutus Cockchafer bettles. Defoliate the vines of grapes at night. Can be controlled by light traps.

Advantages of crop rotations More production of crops is obtained. Soil fertility is not depleted. Less expenses are incurred in crop production. Man and animal labour is fully utilised. Crops are prevented from the onslaught of insect-pests and diseases attack.

Advantages of green manuring Increase of yield after green manuring is usually of the order of 30 to 50 per cent or so. The fertilizing value of the legume crop can be increased a great deal by manuring it with superphosphate.

Advent of cooperative system It is basically started to ameliorate the farmers' or industrial workers' economic conditions. The principle of cooperation can be used in a variety of occupations, may be of agricultural or industrial nature.

Advection The horizontal movement of an air mass across the earth's surface.

Advertising A suitable market can be adapted by advertising agencies.

Aegeratum conzoides Leaf curl is a common virus disease of tobacco of which *Aegeratum cozoides* is an alternate host. The vector is the whitefly, *Bemisia gossypiperda.*

***Aegle marmelos* Beal** A good material for making juice. Family Rutaceae.

***Aegle marmelos* Glutinosa** A good root stock for some Citrus fruits. Fam. Rutaceae.

Aerial, overhead or sprinkler irrigation This method is eminently suitable for areas having uneven topography and where erosion hazards are great. It ensures uniform distribution of water, is adaptable to most kinds of soils and offers no hindrance to the use of farm implements.

Aerobic process of compositing Waste organic materials are usually recommended for making compost. For this, one method depends on aerobic and the other on anaerobic decomposition. In both cases, the farm wastes have to be used as bedding for cattle in order to absorb a large portion of the animal urine. In the aerobic process, the used bedding, the sweepings from cattle-shed and some urine earth from the stable floor are removed every day, mixed with a little cattle dung, and two or three handfuls of wood ash and deposited on a well-drained site to gradually build up a low pile, 12 to 18 inches in height, about 16 feet in width and of any convenient length. The pile is built up before the start of the rainy season. After the first heavy showers, the wetted material in a four feet strip on each side of the long heap is turned with a rake on to the eight feet wide strip in the middle, so raising the height of the heap to three or four feet. This prevents rapid loss of moisture and ensures a quick start of decomposition. When the heap sinks appreciably, which takes about three to four weeks, it is given a turning and made into a fresh heap, thus mixing outside material with that from inside. After about a month or more, depending on the incidence of rains, the heap is given a final turning on a cloudy or moderately rainy day and rebuilt in the vacant part of the original position. The compost becomes ready for use in about four months. If urine earth and cattle dung are not available, the raw organic materials can still be composted, provided more than one-third of the residues is soft and finally broken, such as fallen leaves, lead-mould, kitchen wastes, grass clippings, green succulent weeds, plant trimmings and chopped wheat, barley and soft cereal straw. The use of ordinary soil and wood ashes or lime is, however, essential.

After sowing application of manures In fertile, soils pre-application of at sowing application, manuring is not necessary because soil does have enough manure for preliminary growth of plants. Later, however, manure may be applied according to the requirement and condition of the crop.

***Agave* spp.** A strong living plant used as fence for orchards and fields.

Agmark It is to mark a certain degree of standards used in marketing.

Agmark label For improving marketing, Government of India passed the Agricultural Produce (grading and marketing) Act, 1937. Growers or others willing to observe the rules and to submit to inspection may be granted by the marketing adviser to Government of India the right to use the designation "AGMARK" on their produce and in advertising it also.

Agricultural biotechnology Much of the earliest development in biotechnology research began in agriculture. The emphasis lies in different sectors, depending on support structures and policies perused in past. In some countries, biotechnology has been used to strengthen the traditional production systems, in some others it has been seen as a potential source of ideas for diversifying agricultural production.

Agricultural and Horticultural Society of India In 1820, William Carey, a missionary founded the Agricultural and Horticultural Society of India and four years later he published the first edition of Roxburg's monumental Flora Indica.

Agricultural cooperation In a cooperative organisation, people unite on terms of equality, which finds expression in the adoption of the principle, 'one man one vote' in the constitution and working of a cooperative institution irrespective of amount of capital invested by each member.

Agricultural credit A large majority of farmers, who are left with very little or no surplus at all from their earnings for further investment, require credit for various purposes. Their requirements were met to a very large extent in the past by moneylenders who charged exorbitant interests. Under the Cooperative Credit Societies Act, a system of State loans was introduced to provide for the short and medium-term needs of the agriculturists.

Agricultural engineering Agricultural engineering as defined by International Institutions consists of four main branches, namely (1) Farm implements and power, (2) Rural structures, (3) Soil conservation, drainage and irrigation, and (4) Rural electricity.

Agricultural labour The minimum Wages Act, 1948 is applicable inter alia to employment in agriculture. This Act was amended in 1951, 1954 and 1960. The Act empowers the state to fix minimum wage for various categories of agricultural labourers. Implementation of various provisions of the minimum Wages Act, 1948.

Agricultural marketing Many measures, such as Agricultural produce of Market Act, Grading and Marketing Act etc., for the marketing of agricultural produce in a more orderly manner beneficial to the producers/cultivators are passed from time to time. These measures when adopted protect the farmers from undue exploitation by middlemen and the merchants.

Agricultural produce act The improvement of marketing was greatly encouraged by the passing of the Agricultural Produce (Grading and Marketing) Act of 1937. Under this Act, standards have been established for a number of fruits and fruit products.

Agronomic practices and soil conservation Nature protects soils against wind and water erosion by establishing proper vegetation covers for different land types. The basis of all soil conservation measures proposed to be adopted by man should be patterned as closely as possible after Nature's own way of conservation. Greatest emphasis should, therefore, be laid on the biological methods of erosion-control. Use of land according to its capabilities is largely a matter of crop adaptation and management. Management practices which contribute to the conservation and productivity of agricultural land go to form what is known as conservation farming.

Agronomy The science of managing the land. It includes the theoretical background and practical aspects of growing crops, maintaining the soil's fertility, and the rearing and care of farm animals.

Agro-processing For a scientific marketing of agricultural commodities, certain norms for maintenance of their quality in the market are to be laid down.

Agrostological measures Grass in the same-way as legumes in controlling the soil and water loss. Strips of grass laid in-between those growing field crops are highly beneficial in conserving soil and moisture, and improving the structure and productive capacity of the soil. In places where erosion control by agricultural practices alone or in combination with mechanical means is not possible and where the land is very badly eroded, the best ways to stop any such destruction of the soil lies in the area being put under grass. Grass strips are of special importance in areas where *kharif* rains are not enough to grow legumes for erosion control.

Air The mixture of gases that surrounds the Earth. The composition of dry air, by volume is:

Nitrogen	78.08%	Neon	0.0018%
Oxygen	20.95%	Helium	0.0005%
Argon	0.93%	Krypton	0.0001%
Carbon dioxide	0.03%	Xenon	0.00001%

Air also contains a variable amount of water vapour, as well as dust, pollen, and small amount of other gases.

Air blast The air blast of the seed processing machine should be adjusted to remove a few good seed. This ensures the removal of all light material. In cleaning seed which contain heavy contaminants such as small stones, processors sometimes peed up the bottom fan until all the good seed are lifted up and discharged through the lower air fittings spout. Close adjustment of the lower air allows stones tramp metal etc., to drop through the air blast and discharge out of the good seed spout. When this change of flow, the processor can perform a stoner or gravitation separation on his air-screen cleaner.

Air layering This method is very much similar to layering, the difference is that it is done on the aerial parts of plants or branches of suitable size of trees also. This is also called Gootee or Marcotage.

Air screen cleaner This type of cleaning of seed uses three cleaning principles: aspirator, scalping screens and grading (lower screens). These cleaners vary in size from small two screen farm cleaners to giant commercial machines with upto eight screens and three air-blasts. They can be fitted with any combination of over 200 perforation sizes in woven wire of perforated sheet metal screens.

Air space Space between soil particles or plant cells filled with air.

Air system Any movement of air that is constant in direction and of reasonably long duration *e.g.* the westerly air stream. It is more marked at higher levels in the atmosphere where surface disturbance do not occur.

Alcohol It is produced from rice and other grains by combination of a mould such as Mucor and a Yeast. Starch is being converted into sugar and alcohol.

Aleuranthes busattus, A. wagbini **and** ***A. spiniferus*** These all belong to the group of white fly insects, these are black, orange or white in colour. The larvae of these are smoky colour. They attack many fruit crops.

Alfa-alfa Also called Lucerne. A green clover like plant grown as a fodder crop for cattle and sheep. Deep roots enable the plant to produce foliage even in light porous soils. High yields over a period of five to eight years after planting make this an ideal forage crop.

Alhagi camelorum Fam. Luguminosae. Thorny perennial weed. Control same as for *M. alba.*

Alkali tolerance The presence of large amounts of salts dissolved in water and brought into contact with a plant cell will cause a shrinkage of the protoplasmic lining. This action, called plalmolysis, increases with the concentration of the solution until the plant finally dies. The phenomenon is due to the osmotic movement of water, which passes from the cell towards the more concentrated soil solution. The nature of the salts, the species and even the individuality of the plant, as well as other factors determine the exact concentration at which the individual succumbs.

Almond Prunus (Amyglalus) communis.

Alphonso One of the best varieties of mango in India.

Alternate bearing The tendency of a fruit tree bearing a heavy crop in a year and very light or no crop next year as seen in mango.

Amaranthus polygamous Same as *A. spinosus.*

Amaranthus spinosus Fam. Amaranthaceae. Annual summer season weed. 2,4-D amine 1b/acre/80 gallons of water in six week old cereals gives a good control. MCPB 12 ozs.a.e./80 gallons water/ and linseed in 6 weeks old peas/controls the weed.

Amarantnus viridis Fam. Amaranthaceae. Same-as-above.

Amiben A herbicide. 3-amino-2,5-dichlorobenzonic acid.

Amitrole A herbicide. 3-amino-1,2-triazole.

Ammonification *See* Soil Science.

Amras and amsat It is a dried preparation of mango juice and is a dried product.

Anaerobic process of composting *See* aerobic process.

Analysis of agricultural accounts A careful examination and detailed analysis of accounts at the end of the year is very helpful to know the strength and weakness of the farm business.

Analysis of variance The test of significance of the difference in paired samples can also be put in another and a more instructive form.

Anemometer An instrument that measures the velocity and often direction of the wind. The most common is the cup anemometer. Others include the pressure tube anemometer, the hot-wire anemometer, and the anemometer.

Aneroid barometer An instrument for measuring atmospheric pressure. It consists of a metal box from which the air virtually exhausted. The sides are flexible and expand, and contract with changes in air pressure. These movements are amplified and register on a needle on a calibrated circular dial.

Animal manures Such manures which are prepared from the various materials obtained from animal source such as, wool-manure, bone-meal, dung and urine.

Animal pests These are the common animals damaging crops such as cattle, sheep, goat, forest animals and birds etc.

Annuals Such plants who have their life for one year or season only.

Anona This genus belongs to Anonaceae family. The family has some important edible fruits.

Anona cheriomelia Cheiymoya is one of the best fruits world over. It is the important tropical fruit.

Anona muricate Called Guanabena. Grows a little in Assam (India) and Myanmar.

Anona reticulata *Ramphal*, custard apple or sweet sop. The fruit is slightly larger than *A. squamosa*. The carpels are found to form rind, contains few seeds.

Anona squamosa Sugar apple or sweet sop.

Aonla *Phyllanthus emblica.*, Fam. Euphorbiaceae. Also known by the names of emblic and myrobalon.

Aonidiella auractis, A. citrina* (Yellow scale) and *A. orientalis These all are the scale insects of citrus plants. Damage is done by sucking leaves and very important for Indian condition.

Anthracnose of mango Caused by *Colletotrichum aloeosporiodes*. It attacks leaves and fruits take different shape and colour.

Apocynaceae The family which belongs to the important genus *Carissa*.

Apple A temperate fruit of world wide importance. *Pyrus malus*, Fam. *Rosaceae*.

Apple categories Apple varieties fall into two categories: diploids and triploids. Diploids have plenty of good pollen and are self-fruitful. Triploids are self-unfruitful and become productive only when pollinated by suitable polleniser varieties.

Apple fly speck The fungus *Leotothrium pomi* causes this disease along with sooty mould. The appearance of the fruits is downgraded.

Apple propagation Generally apples are propagated by methods of budding and grafting. Also propagated by methods of ground and trench layering.

Apple pruning 'Modified leader' or 'delayed open centre' tree is preferred to 'central leader' and 'open centre' tree. For 'modified leader' form, cut the yearling whip at about 75 cm, from ground level at the time of planting. During the first summer, allow three to five main branches desired for the framework to grow and pinch the rest. At the first dormant pruning, head back the scaffold branches at roughly about 30 to 60 cm. From the trunk, at a height at which secondary branches are desired. Leader branches are only slightly tipped. During the second summer, remove new growths arising from scaffold branches, and encourage an additional shoot or two to fill up any fancy or to develop a well-balanced tree. During the second dormant pruning, thin out superfluous, interfering and misplaced branches. Head back and branch that shows a tendency to outgrow. At the third dormant pruning, suppress the growth of the leader and allow it to develop as other branches. Remove all interfering branches, water sprouts and sucker growth, allowing only four to six scaffold branches to remain. The tree gives the appearance of a vase in which branches are not allowed to develop closely.

Apple pruning in old trees In bearing trees, the primary consideration in pruning is to maintain a proper balance between vegetative growth and fruit production. A proper balance is reached when mature trees make about 25 cm. of extension growth every year in increasing the bearing area. Mature apple trees require light thinning of branches annually to keep the centre open taught and sun. Old trees, bearing heavy crops of small-sized fruits and making short extension growths require to be pruned more severely to encourage new growth and to maintain vigour of fruit spurs.

Apple root borer *See* Root borer Entomology.

Apple sooty bloch *See* fly speck.

Apple stem bloch Disease caused by *Eurticum salmonicolose*.

Apple storage Can be stored for longer time than the most other fruits can be regularly kept in temperate climate through winter and well with cold storage into the following summer.

Apple varieties Red Delicious, Golden Delicious, Worcester Permian, Newton Wonder (all diploids), Cox's Orange Pippin (triploid), King of Pippins (no. 13), Starking (Royal) Delicious- and Richard.

Application of manures at time of sowing In areas with high rainfall manures should be given at the time of sowing. This is important especially for steep slopes and sandy soils.

Application of FYM Partially rotten farmyard manure should generally be applied to the soil from three to four weeks prior to the sowing of the crop. Given sufficient moisture in the soil, this will allow enough time for its decomposition and improve soil structure. Application too long before sowing season will either cause a drying up of the rotten manure or a too quick decomposition, depending on the incidence of rains, but in either case it will cause a serious loss of ammonia and nitrogen.

Application with irrigation water Manures that are soluble in water are given with irrigation water which spread with water in the field. Unleveled fields should not be fertilized in this way as the spread of water is not smooth in such land.

Applied climatology The use of the science of climatology in the prediction (forecasting) of future conditions, both long and short term. This is of particular importance in agricultural planning.

Apricot *Prunus armeniaca*. Fam. *Rosaceae*.

Aqualin Acrolein or acrylaldehyde. A herbicide.

Aquatic Pertaining to water.

Arbour system Training of grape vines. Sometimes vines are trained as arbours which may be five or six feet high or somewhat higher or lower from the ground. The number of vines per hectare under this method varies from 100 to 500.

Arecanut (*Areca catechuy* L.) Arecanut or betelnut palm is grown extensively in tropical climates. It is a delicate tropical palm, attaining a height of 70 to 80 feet and living from 60 to 100 years. Arecanut is largely used as a masticatory, being chewed with betel (Piper betel) leaves and a little slaked lime. It is said to strengthen gums and help digestion.

Area and production of fruits in India in the year 1994

Name of Fruits	*Area in Hectares*	*Production in Metric Tonnes*
Banana	431684	11900824
Mango	1217362	10118506
Citrus species	433828	3911830
Apples	204996	1298326
Guava	118015	1278883
Papaya	53907	1266185
Pineapple	62225	1006669
Litchi	55498	813388
Grape	38841	70256
Sapota	34832	451142
Total	2653184	3278593
Total with other fruits	3626184	39478593

Arid Denoting any climate or region in which the rainfall is sufficient to support vegetation. It is sometimes defined as an area having an average annual rainfall of less than 250 mm.

Artificial farm manure The straw as it is stacked, or packed into containers, is treated, layer by layer, with a mixture of fertilizer salts of which nitrate of soda or sulphate of ammonia is prominent. Ground limestone or superphosphate are also used. The straw is kept thoroughly moist and compacted, and if the water is warm, decay begins immediately.

Artificial mulch In this case, a foreign material is merely spread over the soil surface. For this manure, straw leaves and other litter may be used.

Artocarpus communis A species more used worldwide and is a staple food in Polynesia.

Artocarpus integrifolia Called Jake, Jack or Jack fruit. Fam. *Moraceae*, (Urticaceae).

Artocarpus lachoocha Monkey Jack. Its both pulp and seeds are edible.

Asphal Bengali name of Litchi.

Aspirators for seed cleaning The use of air to separate seed from other materials is an important principle of many machines, including other separators. Both the aspirator and pneumatic separators use the weight or, more specifically, the terminal velocity of seep to affect a separation.

Assessment of farm business Crop yield per acre, cropped area per farm worker, cropped area per pair of work-animal, gross and net income per worker, number of working days per year for both men and animals, and man-days and animal-days per acre are some of the items that are very helpful in the diagnosis of the strong and weak points of the successful farm business. A well-balanced farm business is the goal for maintaining proper farm accounts.

Asexual Also called vegetative. Reproduce new plants by use of vegetative parts only.

Asexual row In cross pollinated crops like Pearl millet, such rows are maintained for further use in the crossing programme.

Atmosphere *See* soil science.

Attach To join two plant parts of the same type to each other.

Attached method of grafting In this case, the two branches of the two plants are not separated.

Attaching of labels At the time when seeds are placed into bags, a label must be placed on each bag to maintain positive identify of the seed. When bags are closed by a bag-sewing machine, a tag or label can be sewn to the bag. When the filling valves of polythene valve-pack bags are heat-sealed, a plastic or polythene tag can be heat-sealed into the valve.

Augmenting crops These are such crops which help increase in the production of the main crops *e.g.* sowing of *berseem* with mustard.

Aulosira fertilissima It is blue-green algae of particular interest because it possesses a capacity of solubilizing insoluble phosphates and therefore is in much use as soil inocculants.

Autumn The season of the year between summer and winter. In the Northern hemisphere, it is defined as extending from about 21 September the autumnal equinox to about 22 December, the winter solstice from about 2 March to about 21 June. More commonly autumn is understood to indicate the months of September, October and November in the Northern hemisphere.

Auxins These are growth promoting hormones. Their use in increasing fruit size and also to make a graft more successful has been reported. Some auxins bring about seedlessness in fruits.

Avena fatua Fam. Gramineae. An annual weed. Use Dalapon in 10 lbs. to 100 gallons of water plus wetting agent 6 fl.ozs, 100 gallons. Spray during the active growing stages for the grasses. Repeat if necessary on new growths. For wild oats in wheat and barely carbine (Barban) ½–2 lbs. at two leaf stage of weed in very low volume spray may be used to control the weed.

Average composition of farm manure An average farm manure ready for field application can be considered as containing 0.5 per cent of nitrogen, 0.25 per cent of phosphoric acid and 0.5 per cent of potash. Besides the farm manure contains calcium, magnesium, sulphur and all of the trace elements.

Averhoa bilimbi A species of Averohoa fruit which is too sour but is a good material for preservation.

B

Back Cross BC_1 and BC_2 symbols used to designate the first backcross generation and second backcross generation etc. respectively.

Bagger-weigher machine There are many models of this machine available in the market. These are small machines which, when properly mounted beneath a bin, will fill land weigh a bag in a single operation.

Bag sampling of seed For seed lots in bags or other containers, a prescribed sampling intensity should be regarded as the minimum requirement.

Balanced ration for trees It is advisable that proper doses of fertilizers (ration) containing NPK be regularly given to fruit trees for their growth and gainful production.

Ball It is made of soil used for transporting the saplings from the nursery to the planting site.

Bamboo *See* Plant Science.

Banana dehydration If bananas have to be transported through long distance, the fruits are first processed for dehydration to avoid spoilage in transit.

Banana (*Musa paradisiaca* L) Fam. *Musaceae*, Banana occupies a very large area. Though some inferior types of banana are found growing as far north as the Himalaya. Its commercial importance is mainly limited to the more tropical conditions. It is a moist, heat-loving plant and cannot tolerate frost or arid conditions. Cultivated varieties are broadly divided into two groups: table and culinary.

Banana meal Dried and powdered banana is made in Tamil Nadu, India.

Banana propagation Propagation is by suckers or offshoots which sprint at the base of the plant from underground rhizomes. Vigorous suckers with stout base, tapering towards the top and possessing narrow leaves, are selected for planting. The sucker should have a piece of underground stem with a few roots attached to it.

Banana pruning The removal of a large number of suckers in individuals.

Banarsi rai (*Brassica nigra* Koch.) It is garden crop used as a spice. It does not contribute to the supply of mustard oil. The cultivation of white mustard (*Brassica alba* Syn. *alba*) has not ceased in India. *Rai* and yellow *sarson* are self-fertile and the rest of the cruciferous oilseeds, *viz.*, brown sarson, toria, taramira. Banarasi rai and white mustard are self-incompatible.

Band placement Two methods are followed in this method–(1) When crop plants are at a distance of 3 ft. or more from each other, then fertilizer is placed on both sides of each orchard tree. In other case, when in rows there is much less distance among plants, then manure placement is done in rows and is called row-placement. This method is followed in crops like sugarcane, potato, maize and cotton. When there are some plants missing in the row, then such places are left without any supply of manure and are called broken band placement.

Barban 4-chloro-2-butyyl N-(3-chlorophenyle) carbamate, A herbicide.

Barban in wheat and barley Selective control of wild oats in wheat, with barban marks a high point in the development of selectlve herbicides. The point of great interest is that both wild oats and wheat belong to grass family but barban is harmful to wild oats and not to wheat.

Barbed wire These wires are used as fence protection from animals.

Bark graft The scions are cut diagonally as for splice grafting and are inserted between the bark and the wood. This is also called rind or crown grafting

Barley (*Hordeum vulgare* Linn.) This plant provides food for quite a large part of the section of the population. In flour mills sometimes is mixed with wheat for production of *atta* and parched grain is used for preparation of *satu* and for feeding to cattle and horses, and the preparation of 'pearl' barley. One of the grass cereals; its grain is used as a food particularly for cattle, and the straw is either burnt of used in cattle stalls. The grain is also allowed to sprout and then dried in a kiln to make malt, which is used in brewing beer and for making spirits such as whisky.

Barnyard millet (*Echinochloa colana* L. var. *frumentacea*) It is one of the minor millets which are grown largely. The annual area under minor millets is very large. It is quick growing, robust, tufted, annual grass grown in the hot season in a greater part of the world even up to an elevation of 6,500 feet in the Himalayas. The crop is resistant to draught as well as water-logging and is, therefore, cultivated in sands soils and marshy areas.

Barograph A self recording barometer, usually an aneroid barometer. A continuous record (*barogram*) of atmospheric pressure is traced on a moving drum.

Barometer An instrument for measuring atmospheric pressure. The mercury barometer was invented in 1643 by Torricelli. It measures the height of a column of mercury that the atmosphere is able to support in a vertical glass tube.

Barren Under certain circumstances, varieties of figs commonly grown are barren. The achenes fail to develop and the fruit remains dry and eventually fails. The cause of this failure is not known.

Barren land Flat open lands with vegetation limited to small trees and sparse shrubs. The name is applied to the Tundra plains.

Basal dose of manure The amount of manure given to the soil at the time of sowing. This has to be done because if the soil has a meager quantity of nitrogenous manure or the previous crop has been very exhaustive or if in acidic soils the citrate soluble phosphatic manures have to be used or if there is acute shortage of potash in the soil then a potassic manure can be applied at this time.

Basic flow in seed processing As the seed is received in the processing plant, it goes either directly into the cleaning process or into storage to await processing. Drying may be necessary. As processing begins, the first phase (conditioning a pre-cleaning) consists of scalping, de-bearing, hulling, shelling, or any other operation necessary to make the seed flow more readily through cleaners and elevation.

Basic slag This is a by-product of steel factories. Depending on the phosphorous content of the iron ore, it contains from 6 to 20 per cent of phosphoric acid (P_2O_5). Slag from Indian steel mills is poor in P_2O_5 and is not used as a fertilizer. The high grade European slag containing 15 to 18 per cent P_2O_5 is a popular phosphatic fertilizer in central Europe. It is not as soluble as super phosphate but unlike the latter, it is alkaline in reaction and good for acid soils. For effectiveness use, it must be pulverized before application.

Basin cultivation The growing of crops in small hollows or basins, often formed by raising low ridges of soil around a plot of land. The small hollows check rain-wash and collect water for the growing crops. Also *See* Plant Science.

Basin irrigation A means of providing water for cultivation that involves the flooding of basin-like hollows surrounded by earth banks. Also *See* Plant Science.

Baskets The common containers for dispatching fruits from one place to another.

Bats Fruit eating bats are harmful to various fruit, crops are difficult to control.

Bed The land prepared for cultivation. Also known as seed bed. The area in which a river flows is the river-bed.

Bed irrigation This is also called Check irrigation. The small piece of land is bound around by raising a ridge and check the water from flowing out of this bed.

Bees These belong to genus Apis and are very useful insects for man.

Bel *Aegle marmelos*. Fam. Rutaceae.

Belling It is a measure of sugar concentration in a preserve and is described at percentage of sugar *i.e.* 30°C. Belling is 30 per cent sugar solution.

Bending The branches of adjoining trees are bent down and tied to each other. By holding the branches in a more horizontal position than they naturally assume, this forces dormant buds into growth and the crop yield is thereby increased as in guava.

Benefits of green manuring (1) Green manuring of crop helps sustain a suitable amount of organic matter of the soil. (2) Increases the adsorptive capacity of soil. (3) Promote aeration, drainage and granulation of the soil. (4) Increases water holding capacity where needed. (5) In clay it encourages formation of crumb structure. (6) Added organic matter acts as food for microorganisms of the soil. (7) It conserves nutrients. (8) Functions as a cover crop and others.

Benefits of soil drainage Draining the land promotes many conditions favourable to plants and soil organisms. Forces of granulation and aggregation are encouraged. Removal of excess of water also lowers the specific heat of the soil, thus reducing the energy necessary to raise the temperature of the layers thus drained. It is in respect to aeration that the greatest benefits, both directly and indirectly arise from drainage.

Bengal gram (*Cicer arietinum* L.) Gram is the most important and oldest pulse crop grown. Dal, besan, crushed or whole gram, boiled or parched, roasted or cooked, salted or sweet preparations and green foliage and grain as vegetables are the important forms in which it is consumed by the people. Germinated seed is recommended against scurvy. Mallic and oxalic acids collected from green leaves are prescribed for intestinal disorders. Soaked grain and husk are fed to horses and cattle as concentrate and roughage respectively.

Benzoic acid It is used as a preservative at about 0.2 per cent solution.

Ber *Zyzyphus jujube*, Fam. Rhamnaceae.

Bergamot orange From the rind of this, bergamot oil is extracted which is used in perfumes. It is *Citrus bergamia*.

Biased (Sta) *See* expectation of a sample estimate.

Bigrade orange It is the sour, bitter, seville orange. It is grown in Spain and exported to U.K. and other countries for making marmalades. It is also called bitter orange. It is *Citrus aurantium* or *C. vulgarise* or *C. bigaradia.*

Bimodal distribution A distribution which has two peaks in frequency polygon and is not well shaped.

Bioagent A plant or animal used in the control of diseases and insect pests.

Bioclimatology The study of climate in relation to organic life including human beings. This includes the determination of climatic conditions suitable for human habitation.

Biological experiment and fertilizer requirement They indicate relative abundance or scarcity of the different nutrients required by crops from soil, but they give no indication as to the exact quantity of fertilizer that may be applied to make good the deficiency. The dependability of this method can, however, be increased a great deal by coordinating its results with those obtained from field experiments.

Biological methods In this method of weed control a "natural enemy" of the plant is used which is harmless to desired plants. The requirements for the success of this method are (1) the weed species must have been introduced and in the process of introduction must have been freed of its natural predators, (2) natural predators must be introduced to prey upon the weeds but they in turn, must have been freed of parasites in order to carry out their work of destruction unhampered by their natural enemies and (3) the insect must be highly specialised so that it is unable to thrive even under starvation conditions on agricultural plants of the new habit. Root borers, stem borers and internal seed or fruit feeders are more highly specialised than leaf or foliage feeders.

Biomass The weight or volume of living organisms of one animal or plant species per unit area (species biomass), or of all the species in the community (community biomass). The weight or volume of organic matter per unit area. It may be measured in grams of dry matter per square metre, so that comparisons between different environments can be made. Biomass may refer to the total organic content of the biosphere to that of a specific habitat such as a pond or to that of a medium such as soil.

Biotic Denoting living organisms *e.g.* the biotic environment is made up of living components (*i.e.* plants and animals) as distinct from abiotic components such as rocks and air.

Black barriers Also called dewberry, *Rubus longanbacus, R. allagheniesis* and *R. flagellaaens*.

Black cotton *See* Soil Science.

Black gram (*Phaseolus mungo* Roxb.) Black gram is highly prized pulse, very rich in phosphoric acid. It is used more or less in the same way as green gram. It is also used in prepared paper (a kind of wafer) and bari (spiced balls of ground dal).

Bladed harrows These are called bakhars and guntakas. This is a universal implement used in Deccan area. It is one of the best implements of primary tillage. The farmers use it for preparing the land for sowing the seed. It consists of a steel blade of about 1½" to 3" in length and 3" in width. It can cultivate the land to a depth of 3" to 4". Even the dry land tillage is done by means of this implement by adding a wedge to it. If the soil is very dry, a man stands on the implement to give in the weight. It can cover two to four acres of land in a day because of its greater width. The implement is worked cross-wise in the fields to achieve better tilt.

Blanching Before filling the fruits into the cans, they are immersed in boiling water. For this purpose, wire baskets are used so that the whole basket with sizeable quantity of fruits can be immersed at a time.

Bleach To discolour fruits before making preserves.

Blood meal A very rich protein food. It is also an important organic fertilizer. *See* dried blood.

Blossom showers (mango showers) The rain showers that occur from March to May in the monsoon region of S.E. Asia.

Boiling and canning of fruits Wash fruits in cold water to remove dirt and other extraneous matter. For preparing the material, apricots are cut into halves and stones removed; figs are boiled in water for a few minutes till tender, grapes are stemmed; guavas are peeled, cut into halve and seed cavity removed; yellow bulbs of jack fruits are used as such after removing seeds; litchies are shelled and stones removed; mangoes are peeled and cut into slices; papaya is peeled and cut into small slices; peaches are cut into halves, stones removed and dipped in 1 per cent boiling caustic soda solution for ½ to 1 minute to peel them, washed thoroughly in cold water; pears are peeled, cut into halves and cores removed; pineapples are peeled and cut into round slices; and plums are only washed.

Fill the prepared fruit into containers in any convenient manner. Size of containers recommended is A-2½ size cans and one pint jars. Add boiling hot sugar syrup (40 to 55 per cent) to containers. Heat containers in water till temperature in the centre of the container reaches 170°F., and then seal them air-tight, sterilize apricots, jackfruit bulbs, lithcies, mangoes, papayas, peaches and pears for 25 to 30 minutes; figs and grapes for 12 to 15 minutes; guavas and pineapple for 20 to 25 minutes; and plums for 15 to 20 minutes. Remove containers from sterilizer and cool cans in running cold water; and glass jars in air.

Bokhara plum *Prunus bokharensis*, Fam. Rosaceae.

Bone-char It contains 1–2 per cent nitrogen and 20–35 per cent phosphorus.

Bone-meal A desirable supplement for laying stock. Its chemical form is $Ca_3(PO_2)_2$ and, is available in the slaughter houses. It is a good manure for all crops. It should be given in very finely ground condition to get good results. It may be applied in the soil before one month of sowing. It contains 8–12 per cent nitrogen and 20–22 per cent phosphorus.

Bone-meal use The use of raw bones as manure for fruit trees is a time old practice. Burying an animal skeleton under a fruit tree is known to benefit its growth and bearing. Yet a large quantities of bones used to be exported until a few years ago. Exports have since declined and bone-meal (*i.e.* ground bone) is now a widely used phosphate fertilizer. It is available in two forms: (*i*) raw bone-meal, and (*ii*) steamed bone-meal. Steaming of bones under pressure removes fats, greases, nitrogen and glue-making substance. Thus, while raw bone-meal contains 4 per cent slow acting organic-nitrogen and 20–25 per cent insoluble phosphoric acid, steamed bone-meal contains only 1 to 2 per cent nitrogen but 25 to 30 per cent phosphoric acid. Steamed bones are more brittle and can be readily ground. This is an advantage as the rate of availability of phosphoric acid in bones depends largely on their degree of pulverisation. Bone-meal having particles not larger than 3/32 inch in size is considered suitable for use as fertilizer, but the more finely powdered it is, the quicker its P_2O_5 becomes available in soil.

Bone-meal time to apply Being relatively slow acting, bone-meal should not be used as a top dressing; it must be incorporated in the soil in order to become available. It may be applied either at sowing time or a few days before sowing and should be broadcast. It is particularly suitable for acid soils. It is considered a safe manure for all crops.

Bone raw It contains 2–4 per cent nitrogen and 22–25 per cent phosphorus.

Boron It is a micro-nutrient essential for healthy plant growth.

Boron deficiency in cruciferae Boron necrosis developed in fleshy pith of stem and core as plants approach maturity; in turnip, internal breakdown of brown necrotic tissue occurs. Apply 20 lbs. of borax per acre.

Bottling and canning of vegetables Wash vegetables thoroughly in running water. Prepare them as for cooking; dip the prepared vegetables in boiling water for 2 to 5 minutes and then dip in cold water to cool.

Fill the vegetables into containers in any convenient manner. Sizes of containers recommended is the same as for fruits. Add boiling hot solution of common salt (2 per cent). Heat containers till the temperature in the centre of the containers reaches 170°F 175°F and seal the containers air-tight.

Sterilize peas, green gram and potatoes for 40 to 45 minutes; carrots for 30 to 35 minutes; cauliflower for 30 to 40 minutes; French beans for 35 to 40 minutes; ladyfinger for 25 to 30 minute and turnips for 30 to 45 minutes, all at 10 lb. Pressure (240°F), and spinach for 50 minutes at 15 lb. Pressure (252°F), in a pressure-cooker. Remove containers from pressure cooker and cool cans in running cold water, and glass-jars in air. Adjust the pressure required for sterilization according to the altitude.

Bottling of fruits In the home, it involves the use of glass jars with proper lids to seal them air-tight, and can be done with ordinary kitchen equipment. For canning of fruits and vegetables in the home, sanitary tin cans, a can-sealer and a pressure-cooker are required in addition. The method consists of washing and preparing fruits and vegetables, filling the prepared material into containers, adding syrup on brine (common salt solution), exhausting, *i.e.* removing air from containers, sealing containers air-tight and cooling the sterilized containers.

Bract An additional leaf like structure attached to the-base-of flowers.

Branding This term used to identify the product and branding is the practice of the identifying the product or line of product by a special name or brand or symbol.

Brassica compestris A very important winter oilseed crop plant.

Brassica sinensis Family Crucifereae. Annual winter season weed. In cereals 2, 4-D amine at 1 lb. a.e./acre in 80 gallons of water. Cereals should be sprayed after they have fully tillered and prior to boot formation. In peas and linseed MCPB 12 ozs. a.e./ acre 5 to 6 weeks after sowing. These weeds are highly sensitive to 2, 4-D which therefore should not be used in Brassica crops.

Bread fruit *Artocarpus insisa.*

Breeder seed Seed produced by the breeder and used to produce foundation seed.

Bridge grafting This method is under a condition as suppose the bark of tree accidentally peels off due to crash of a vehicle or instrument or like so, then the damaged-portion is replaced by a patch of bark having a twig or bud etc. This can be done also in case the bark has rotten and fallen off, then it is covered by anew bark with a twig or developed bud.

Broad based ridge terrace This consists of a wide, low embankment formed on contour by excavating soil from both sides of it. It is more suited to low rainfall areas and fairly absorptive soils.

Broken band placement *See* band placement.

Brambles Some of them bear fruits of excellent quality and while they do not keep well, it can be made into a very good jam. Wild fruits are plucked and used fresh or made into jam. Propagation is by means of suckers, root cuttings and rooted tips of canes. The plants are put a few feet apart in rows. After bearing, old canes are cut out, and in some cases it is desirable to thin out the suckers.

Brands of fertilizers In order to meet the demand of the well-balanced fertilizers suited to various crops and soils, a large number of brands are being manufactured. Some of these contain all the three fertilizer elements. Some two and some have the single element.

Buchanania latifolia *Chiraunji* in Hindi. It is tree of importance from horticulture point of view. Family Anacardiaceae.

Buckwheat An annual herbaceous plant, native to Asia, the seeds of which are used in Europe for feeding poultry, horses and cattle. In North America, buckwheat flour is the basis for small cakes, which are fried and eaten for breakfast.

Bud The initial compact vegetative living part.

Budding A method when a bud of a scion is implanted on the root stock of another plant of the same species / type.

Bud stick The portion of the branch having bud in it.

Bugs This class of insects are harmful kind as far as crop destruction is concerned.

Bulb crop vegetables The two important bulb crops, onion and garlic belong to the genus Allium of family Liliaceae. They can be grown under a wide range of climatic conditions, but they succeed best in mild season without extremes of heat and cold. They are grown during winter and harvested before the real hot season begins.

Bulbil A underground stem in which a bulb is further divided into small bulbils such as in garlic.

Bulk seed sampling When sampling seed lots in bulk (heaps, bins, wagons etc.) or streams of seed during processing operations, the sampling intensity should be regarded as a minimum requirement for obtaining the bulk sample.

Bullock labour This consists of cost of feeds, maintenance labour, interest and depreciation in the value of bullocks and miscellaneous such as housing, shoeing, medicines etc, less the value of manure produced. In the event of death of the bullock during the year, the entire value minus depreciation already paid, should be included in the bullock labour. The cost of manual labour for looking after the animals may be debited to this item at the rate of one man for ten animals.

Bullock's heart (*Annona recticulata* L.) The bullock's heart is an allied fruit which is rarely cultivated. It prefers a heavier soil than custard apple and is less resistant to cold and frost. It is easy to propagate by inarching. It gives fruits from January to May and the yield goes upto 45 kg per tree.

Bulrush millet (*Pennisetum typhoideum* Pers.) Bajra is one of the most important millet crops. Grain is ground into flour and prepared as mash or unleavened bread. It can also be fried for producing a nutritious food. Plant stalks are fed to cattle or used for thatching purposes. It is suited for warm areas of low rainfall (15 to 20 inches), doing best under conditions of light showers followed by bright sunshine during the growth period. Heavy rains are disastrous to the crop.

Bumper crop A very good crop having a probability of very high yield.

Bumper yield A crop yielding a very high produce.

Bund An artificial quay, dam of embankment or a small ridge between rice fields.

Bunding of the fields on contour lines With proper surplusing arrangements and division of the land into compartments to reduce runoff and to conserve soil and water is the first step. Bunding should as far as possible be done combined with other recommended agronomic practices to obtain the best results by dry farming.

Bund formers For soil conservation, sometimes it is necessary to make temporary low level bunds of the height of bout 7″ to 9″. These bunds last for one agricultural season and are then either washed away or obliterated by ploughing and cultivation.

Butterfly A metamorphic stage in the re-production of insects.

Buy Bullor's law Law postulated by the Dutch Meteorologist Christoph Buys Buller in 1857 which states that if an observer in the N. hemisphere stands with his back to the wind, atmosphere will be lower on his left hand than on his right, in the S. hemisphere pressure is lower on the right. This is the result of the coriolis force on the earth. Expressed in a different way, the law states that in the Northern hemisphere winds move in an anticlockwise direction around centres of low pressure and clockwise around centres of high pressure in the S. hemisphere; the reverse is true also for coriolis force.

By-product A material produced in addition to the main product in any chemical, physical or mechanical process.

C

Cacao (*Theobroma Caco* L.) A tree of equatorial and tropical regions that bears large oval seed pods containing beans from which cacao and chocolate are made. Since chocolate is consumed mainly in temperate lands, cacao is a valuable export for some Third World countries. Cacao is both a beverage and a food crop. It is used for manufacturing chocolates, cacao powder, cacao butter etc. The shells of the beans are used as cattle feed or manures.

Cakes as manure. Cakes of different vegetable origin especially of oilseeds are important organic manures. They are more quickly effective than the dung or compost manures. The main cakes generally used as manures are:

Sl.No.	*Name of Cake*	*Nitrogen %*	*Phosphorus %*	*Potash %*
1.	Groundnut	7.6	2.3	2.2
2.	Mustard	6.5	1.0	1.4
3.	Safflower	5.8	1.3	1.2
4.	Linseed	5.7	1.6	1.6
5.	Sesame	5.0	1.0	1.1
6.	Coconut	3.7	1.9	1.8

Sl.No.	Name of Cake	Nitrogen %	Phosphorus %	Potash %
7.	Castor	5.0	1.8	1.6
8.	Neem (Margosa)	4.4	1.0	1.4
9.	*Karanj*	3.5	0.7	1.3
10.	*Mahua*	2.6	0.8	2.8
11.	Rainchilli	4.5	2.0	1.9

Calamondin *Hazara*, qumquat, *Citrus microcarpa*, Fam. Rutaceae. A native of Philippines.

Calcareous manures These manures have calcium as their major constituent.

Calcium cynamide This is a black coloured fertilizer. It contains about 20 per cent nitrogen.

Calculations of dosage level of herbicides The quantity of the toxic ingredient to which the weed-killing properly of the chemical is due, is different in different formulation of herbicides. The active ingredient is that part of the chemical formulation which is directly responsible for the herbicidal effects. In some herbicides, the entire molecule may be active unit. If the chemical is 90 per cent pure, it is considered to have 99 per cent active ingredient.

Calibration of resistance blocks Resistance blocks are relatively cheaper and simpler to operate than tension-meters. Both required to be calibrated in terms of available moisture for a particular soil. Both the methods are affected by the diurnal change in temperature. A number of improved types of tension-meters and resistance block have been designed but none of them are perfect in all respect.

Calotropis gigantea Fam. Asclepiadaceae. It is a large shrub if allowed to grow uninterrupted. It can be controlled by use of 2,4-Damine 1–2 lbs. a.e./acre before its flowering.

Cambrian The earliest geological period of the Palaeozoic era and the system of marine rocks that were laid down during this period. The Cambrian rocks contain the first shelled fossil remains and are the earliest rocks for which fossils can be used for dating and correlation trilobites, brachiopods and gastropods were abundant.

Campbell Stock's recorder An instrument for measuring and recording the duration of bright sunshine. As the Sun moves round during the day, a spherical lens focuses its rays to burn a trace on a recording card. The length of the trace indicates the sunshine duration. The instrument has limitation that it only records bright sunshine early morning and late evening sunshine may not be detected.

Canal A man-made water way used to transport goods or irrigation water. Canals may be built to link inland towns or cities to the coast. Canals are particularly useful for transporting bulky goods when the time taken in transit is not highly significant.

Candied fruit *See* Crystallised fruit.

Cane killing weed of sugarcane (*Saccharum officinarum*) A weed growing in clusters near base of clump small lanceolate leaves and whitish-pink terminal flowers, white to light pink. The parasite is the *Striga* spp. practise crop rotation with legumes; apply seedicides (Methoxone, Agroxone and Paraxone).

Cane sugar *See* Sugarcane. *See* Sucrose.

Cane system of grape vine training This system is suitable for training on two-wire trellis. The trunk is headed back as in the head system; four arms, two on each side of the trunk, are allowed to develop and pruned to 30 m. length after a year. Fruit canes, each carrying 10 to 20 buds, develop on each arm which are tied to the trellis wire. A renewal spur is left on each arm just below the cane for next year's crop.

Canning of fruits and vegetables Important method of preservation using the term to denote preservation either in tin cans, glass jars or bottles.

Canning of curried vegetables Canning of curried vegetables is similar to canning of ordinary vegetables. Instead of common solution, a spiced gravy of the desired taste is added to the containers. The material is sterilized 5 to 10 minutes longer. A gravy sufficient for one dozen cans may be prepared of the following ingredients: mustard ¾ oz., coriander powder ¾ oz., red chilli powder 1½ to 1¾ oz., common salt 3 oz., vegetable fat 14 oz. and water sufficient for 12 cans. Fry mustard seeds in fat till they crack, add other ingredients, continue frying for a few minutes more, add required quantity of water and boil. Freshly extracted fruit juice deteriorates rapidly. It is usually preserved as such in pure form or as squash, cordial, syrup, sherbet, barley water etc.

Cannabis sativa Fam. Cannabinaceae (Urticeae) Annual dioceious weed in nature. Susceptible to 2,4-D ester or a mixture of 2,4-D and 2, 4, 5-T at ½ lbs. a.e./acre during flowering state.

Capital goods Goods that will be used in the manufacture of other goods, *e.g.* looms are capital goods that are used to manufacture textiles. Raw materials such as wood and china clay are used to produce and are also capital goods since they are used to produce such things as paper and porcelain.

Caprification A special type of pollination in figs (Caprifigs) brought about by a small insect-wasp.

Caprifig Figs of primitive type which produce pollens. This is used for pollination by the wasp insect to cross-pollinate other varieties which do not produce viable or any pollen.

Capsicum species These are important vegetable and spices plants.

Carambola *Averrhoa carambola*, Fam. Oxalliaceae.

Carbon cycle *See* Soil Science.

Carbyne Same as Barban.

Cardamon (*Elettaria cardamomum* Maton) Cardamon is an important spice used for flavouring or medicinal and masticatory purposes. The seeds contain 2 to 8 per cent of strongly aromatic volatile oil with a pleasant cooling effect.

Care of farm machinery The farm machinery should be maintained for its long economic life.

Care of ploughs Daily checking of plough adjustment tightening of bolts and nuts on the frame and bottoms and the regular attention to lubrication and replacement of worn out parts prolong the life of the ploughs and lessen the cost of ploughing.

Carica papaya It is an important fruit crop. The leaf curl caused by virus is a very serious disease of papaya. They also attach other crops like beans, sann hemp, cotton, bhindi (*Hibiscus esculentus*), brinjals, tomatoes etc.

Caricaxanthin Papaya is the second best source of vitamin A after mango. While this vitamin is generally associated with carotene, the yellow pigment in the papaya is the caricaxanthin.

Carnot cycle The idealized reversible cycle of four operations occurring in a perfect heat engine. These are the successive adiabatic compression, isothermal expansion, adiabatic expansion and isothermal compression of the working substance. The cycle returns to its initial pressure, volume and temperature, and transfers energy to or from mechanical work. The efficiency of the Carnot cycle is the maximum attainable in a heat engine. *See* Carnot's principle.

Carnot's principle (Carnot theorem) The efficiency of any heat engine cannot be greater than of a reversible heat engine operating over the same temperature range. It follows directly from the second law of thermodynamics, and means that all reversible heat engines have the same efficiency independent of the working substance. If heat is absorbed at temperature T and given out at T_2, then the Carnol efficiency is $(T_1-T_2)/T_1$.

Carotene A carotenoid pigment, examples being lycopene and A and B carotene. The latter compounds are important in animal diets as precursor of vitamin A. *See* carotenpids, photosynthetic pigments.

Carotene content of mango It contains more carotine than any of the other fruits and compares well with many of the leafy vegetables.

Cartenoids A group of yellow, orange of red pigments comprising the carotenes and xanthoplylls. They are found in all photosynthetic organisms, where they function mainly as accessory pigments in photosynthesis and in soma animal structures *e.g.* feathers. They contribute, with anthocyanins, to the autumn colours of leaves since the green pigment chlorophyll, which normally masks the carotenoids, breaks down first. They are also found in some flowers and fruits *e.g.* tomato.

Carrying capacity of land The ability of an area of land to support people or animals. The capacity can be measured in terms of the unit of land that is required to support one person or one animal, for example, it has been calculated that the alluvial deltas of SE Asia, which are used for growing rice, have a carrying capacity of 500 people per square kilometre. By comparison, the semiarid region of W. Africa may only have a capacity of one person to 10 square kilometres.

Carthamus oxycantha An annual weed, 2-4-D amine, 1b. a.e./acre/ 80 gallons of water in six weeks old cereals. Do not use 2,4-D in dicot crops. For this weed plant, in gram and Brassicas, 2,4-D 2 lbs. a.e/acre immediately after harvesting prevents and seed setting in lateral branches of this weed.

Cash crop A crop that is primarily grown for sale and not for use by the grower and his family.

Cashew (*Anacardium occidentale*) Fam. Anacardiaceae. Cashew is grown both for its fruit (cashew apple) as well as nuts, mainly in Peninsular India particularly along the coast. Cashew does not tolerate severe summer or winter conditions. It does not grow satisfactorily at elevations of soil conditions. It wants a free drainage soil. It grows in areas having rainfall varying from 50 to 400 cm.

Castor cake It contains 4–5 per cent nitrogen, 1.8 per cent phosphorus and 1.3 per cent potash.

Catch crop A crop that is either grown when the ground is fallow between the harvesting of one crop and the planting of the next, or grown in between the rows of an existing crop. Vegetables are sometimes grown between rows of olive trees in Mediterranean lands.

Cattle trespass In order to prevent the depredation of crops by stray cattle, the Government of India enacted legislation as early as 1871, for regulating the trespass of cattle by a comprehensive Act known as the Cattle Trespass Act.

Cattle urine manure This is an important organic manure and feed of cattle gives out nitrogen in various excreta such 30 per cent in dung, 50 per cent in urine. So its addition to the manure pit adds sizeable quantity of nitrogenous manure to the manure.

Cattley guava *Psidium cattleianum*. Fam. Myrtaceae.

Caution for mixing fertilizers Ammonium nitrate is an explosive chemical and, therefore, dangerous for home mixing. Mixtures containing other nitrates should be made only in such quantities as are to be used immediately, because they absorb water quickly and are not suitable for storage. Bone-meal, sulphate of potash and muriate of potash can be-mixed with all fertilizers.

cdl *See* critical day length.

CDAA 2-chloro-N, N-diallylacetamide, a herbicide.

CDEC 2-chloroallyl diethyl dithio carbamate, a herbicide.

Ceiling on holding Legislation for ceiling on existing holdings has been enacted in all states and Union Territories.

Cell *See* Plant Science.

Celsius scale A temperature scale in which the-temperature of melting pure ice is taken as 0° and the temperature of boiling water 100°C (both at standard pressure). The degree Celsius (°C) is equal to the kelvin. This was known as the centigrade scale until 1948, when, the present name became official.

Cenchrus ciliaris A good fodder grass.

Central leader When a tree is trained as to have a central branch the longest one, smaller ones growing out from it in various directions. Such a tree is physically strong but as an orchard, tree has some disadvantages. It is also called closed central system.

Central land mortgage banks The long-term needs of the farmers were met by cooperative central land mortgage banks. These banks usually obtain the bulk of their finance by floating debentures. The state governments generally guaranteed the principal and interest on, the debentures.

Central state warehousing corporations The main functions of the Corporation are to acquire or build godowns and warehouses, to run warehouses for the storage of agricultural produce, to subscribe to the share capital of the state warehousing corporations and to act as an agent to the government for the purchase, sale, storage and distribution of agricultural produce.

Centrifugal pump A device commonly used for transporting fluids around a chemical plant. Centrifugal pumps usually have 6–12 blades rotating inside a fixed circular casing. As the blades rotate, the fluid is imparted out of the pump along a pipe. Centrifugal pumps do not produce high pressures but they have the advantage of being relative cheap because they are simple in design, have not valves and work at high speeds. In addition, they are not damaged if a blockage develops. (Compare displacement pump).

Centripetal drainage A pattern of drainage in which the streams flow towards a central point.

Cereals These food grain crops include wheat, barley, rice. A cultivated grass used as human or animal food. *See* Plant Science.

Cereal fodder crops The principal cereals grown as fodders are Andropogon, sorghum (jowar), *Zea mays* (maize), *Pennisetum typhoideum* (bajra), *Setaria italica* (kangani), *Panicum miliaceum* (cheena), *Eleusine coracana* (ragi), *Avena sativa* (oats).

Certified seed Seed used for commercial crop production produced from foundation, registered, certified seed and certified by an official certifying agency.

Change in bulk of farm manure Due to loss of carbon dioxide, and water during the decay processes, there occurs a considerable change in the bulk of the manure. Fresh excreta losses from 20 to 40 per cent in bulk by partial rotting and 50 per cent by becoming more thoroughly decomposed.

Change in fertilizer schedule Changes in one or more of them are bound to influence the economics of fertilizer use. The use of fertilizers is also subject to the 'law of diminishing returns.' This means that the rate or increase in crop yield decreases after a certain point is reached regarding the quantity of fertilizer used and consequently the value of the additional yield finally.

Channel terrace It consists of a wide and relatively shallow channel excavated at suitable intervals on a failing contour with a suitable longitudinal gradient. The excavated soil is deposited as a wide low ridge along the lower edge of the channel. This is recommended for relatively impervious soil and in heavy rainfall tracts.

Characteristic of farm manure (1) It is moist condition of manure, (2) Its variability, (3) Its low analysis, (4) Its low grade, (5) Its unbalanced nutrient condition, (6) Its unbalanced constituents, (7) Its residual influence exerted in the system and (8) Its rapid fermentation processes.

Characteristic of a good fodder (1) It should be palatable, digestible and non-injurious at the state at which it is to be fed, (2) It should occupy the land (for one crop season) for as short a time as possible more so when grown under irrigation, (3) It should yield as large a crop of fodder per acre as possible, (4) It should need a little tillage and manuring, even though good fodder do as a rule need fairly high tillage and manuring, (5) It should be capable of being preserved, generally as silage or hay, without any deterioration.

Charred bones as manures In some cases, charred and powdered bones are used as manure. Charring destroys about half the nitrogen but leaves intact practically the whole of P_2O_5 in a quickly available form.

Charimoya *Annona charimolia*. Fam. Anonaceae.

Cheekoo *Achrus sapota*.

Cherry (*prunus avium* L.) Cherries are of two types: sweet, used for desert and sour, used for cooking purposes. Grown mainly in the Shimla Hills, Kullu Valley and Kashmir at elevations above 1,500 metres. It is desirable to choose varieties that will ripen in succession in order to obtain the crop over a longer period. A large number of varieties are self-unfruitful and do not set fruits with their own pollen. As they area also cross incompatible, only the compatible varieties, whose period of flowering overlaps to effect cross pollination, should be interplanted to get commercial crops.

Check irrigation Same as bed irrigation.

Chemical analysis of dry animal manure It is extremely difficult to quote reliable average figures on mixed farm manures as it applies to the land. This is because of a number of variable factors enter in which may change the amounts and proportions of nitrogen, phosphoric acid and potassium present.

Chemical analysis of farm manure In spite of the marked variability of farm manure, representative composition figures are generally suggested. The average farm manure ready for field application can be considered as having 0.5 per cent nitrogen; 0.25 per cent phosphoric acid and 0.5 per cent of potash.

Chemical analysis and fertilizer requirement Chemical analysis of soils or of plants growing on the constitutes the modern method of determining the fertility requirement status of soil Such analysis provide information on fertilizer requirement of a crop also.

Chemical fertilizers These are prepared by various chemical methods. Ammonium sulphate, ammonium nitrate, potassium nitrate, potassium chloride and superphosphate come in this category.

Chemical analysis for fertility needs Two general types of analyses are done–(1) total, and (2) partial. In the total analysis, the total amount of any particular constituent present in the soil is determined. In partial analysis, which are at many kinds, but one type which shows much promise is not to determine the soil solution but of replacement from an absorbed condition.

Chemical analysis of synthesised fertilizers It is easily done in the laboratory and its interpretation is easy. Generally the labels indicate the three main ingredients as percentages of N, P_2O_5 and K_2O present in any bag.

Chemical method of soil reclamation It is done by use of calcareous amendments. *See* reclamation.

Chemical method of weed control Controlling weeds in crops through the use of weed-killers leads to increased yields and ensures efficient utilisation of irrigation, fertilization and other plant protection measures such as insecticidal and fungicidal sprays.

Chemical preservatives Salt acts both as by osmosis and as a poison, while vinegar acts only as a poison. A product must contain about 65 per cent of sugar, to be preserved, while 15 per cent of salt or 1 per cent of acetic acid is ordinarily effective.

Chemical treatments of cuttings In many cases, chemical treatment has increased the percentage of cuttings forming roots, but no one treatment has provided uniformly helpful. Potassium permanganate, manganese sulphate, acid and sugar solutions have all seemed to give good results with certain species under certain conditions.

Chenopodium album Fam. Chenopodiaceae. An annual weed, 2, 4-D amine ½–1 lb. a.e./80 gallons of water in monocot craps; in peas and linseed MCPB 12 ozs. a.e./acre/80 gallons, 5–6 weeks after sowing. A safe procedure to avoid injury to the crop plants.

Chenopodium murale Chenopodiums are winter season broad leaves weeds. Same as for *C. album.*

Cherimoyer One of the best fruits of genus Annona and is named *A. cherimola.*

Cherry These are two types of fruits belonging to genus *Amygdalus* among which the sweet is *A. avium* and the sour type is known as *A. serasus.*

Chestnut *Castanea sativa* and *C. crenata,* Family Fagaceae.

Chicking vetch (*Lathyrus sativus* L.) A papileoneacious plant used as a short duration crop for fodder. Some people use the grain as pulse, particularly in times of famine. The unripe pods are occasionally used as green vegetables.

Chile saltpetre *See* sodium nitrate.

Chinese or sand pear *Pyrus serotina,* Fam. Rossaceae.

Chillies (*Capsicum annuumm* L.; *Capsicum frutescens* L.) Chilli, also called red pepper, is an important condiment crop, grown for its pungent fruits, which are used both green and ripe (the latter in the dried form), to impart impugency to food. It is also used medicinally, and in chutnies and pickles. The pungency is due to active principle Capsicin contained in the skin and septa of the fruit.

Chilli pickle *See* lime pickle.

Chilling To keep products at very low temperatures to avoid bacterial damage.

Chloride of lime It is obtainable in the form of a white powder. One pound of it mixed with 2½ gallons of water makes a very goad deodorant. It may be used for out-houses, baths, etc. but it is not as good as an disinfectant as coal-tar products.

Choice of class intervals (Sta.) A frequency distribution will vary with the number of class intervals. The choice of class intervals therefore requires careful consideration.

***Chorchorus capsularis*-jute** It is an important fibre crop. It suffers from a serious disease, the stem rot caused by *Macrophomina phaseoli.* Both the quality of the fibre and its out-turn are adversely affected.

Chrysopogon montanus A good fodder grass.

Chyme Also food reduced to liquid form.

Clinicus arvensis Same as *Carthamus oxycantha*.

CIPC Isopropyl N (3-chlorophenyl) carbamate, A herbicide.

Circium arvense A perennial weed. Same as for *Carthamus oxycantha*.

Citrangequat The hybrid between citrange and kumquat.

Cimequat Hybrid between lime and qumquat. It bears fruits similar to that of lime and in some cases equally good and is more hardy than the lime.

Citrange In an effort to secure varieties which can be grown in regions subject to frost, the sweet orange has been crossed with the trifoliate orange and the hybrid is the citrange.

Citron Family *Rutaceae, Citrus aurantium* Is a good root stock for propagating other fruit types.

Citrus fruits All such fruits belonging to the family Rutaceae and genus Citrus. Citrus is grown in a large area in the world. The Citrus fruits are grown in almost all kinds of soils varying from heavy black soils to shallow open soils. Some of the varieties of Citrus seem to adapt themselves to soil conditions better than others. They thrive in free draining alluvial or medium black soils of loamy texture. A hard substratum or a sticky impervious layer is very injurious. Soils having a high water table should be avoided. Though Citrus trees on the whole do well in a dry climate having rainfall between 75 and 12 cm., certain species like pummelo and certain mandarin oranges thrive in heavy rainfall areas.

Clean cultivation When there is a single crop in a field and the field is clean off the weeds and any other undesirable plant groups.

Cleft The cleaved end of the rootstock while doing wedge grafting. See wedge grafting. Cleft hoofs are found in cattle.

Climate The four most important climatic requisites of successful agriculture are water, air, sunlight and soil. Variations in incidence of the first three are connected with climate. The climate, therefore, constitutes the most important single factor, which determines the possibility of profitable farming.

Classification of manures This is done on kind of origin, modes of getting them, composition etc. of the manures.

Classification of weeds The classification of weeds (terrestrial as well as aquatic) into annuals, biennials and perennials is based on the time required for the completion of the life-cycle and the mode of reproduction. Annuals live and produce their seeds in a single growing season. Biennials need two years to complete the reproductive cycle. The multiplication of both the annuals and the biennials is through seed. Perennials live indefinitely and are propagated not only by seeds but also often vegetatively through underground structures such as rhizomes, stolons, bulbs and tubers. Perennials are of two types, the simple and the creeping. The former ramify only by seed. They have no normal means of spreading vegetatively. However, if injured or cut, the severed portions produce new plants. Creeping perennials spread by creeping roots, creeping above stems (stolons), creeping below ground system (rhizomes). Some weeds propagate by means of tubers, which are modified rhizomes adapted for food shortage. Aquatic weeds (water inhabitants) are classified into three types, *viz.*, submerged, emerged and floating.

Clay-humus complex The close association between the colloids of clay (minerals) and finely divided particles of humus (organic) in the soil.

Climatology The study of climate. It includes the study of climatic phenomena, their causes and their influence on the natural environment.

Closed central system *See* central leader.

Closed drains As the open drains have the disadvantage of wasting the land, are ineffective and inconvenient, encourage erosion etc., closed or under-drains should be used wherever possible.

Cloud A mass of minute water droplets and/or ice crystals formed by the condensation of water vapour and held in suspension in the atmosphere. Condensation which results from cooling usually takes place around nuclei such as dust, smoke particles and salt. The cooling may be caused by convection, uplift over mountains or ascent in depressions. Clouds may be present at heights ranging from ground level (fog and mist) upto over 13,000 m. several classifications of clouds exist but the one international agreed on and most used is based on cloud appearances and height. It comprises ten principal forms. The low clouds (upto a height of about 2,000 m) which comprise stratus (st), stratocumulus (Sc) and nimbostratus (Na). Above these are the medium clouds (2000–7000m) which comprise altocumulus (Ac) and altostratus (As). The high clouds (7000–13000 m) are cirrus (Ci), cirrostratus (Cs) and cirrocumulus (Cc). Some grow vertically and cannot be classified solely by height, these are cumulus (Cu) and cumulonimbus (Cc).

Cloud seeding The modification of weather by introduction of artificial nuclei into clouds to include the minute droplets of water or particles of ice to coalesce and form raindrops, or ice pellets. The idea was first put forward in the late 19th century but experiments began only in 1940s.

Cockchafer beetles *See Adoretus lesiopygus.*

Coconut (*Cocos nucifera* L.) Unlike the other oilseed crops, the coconut is a perennial tree grown as a permanent garden or plantation crop. It is grown extensively in the numerous islands of tropical seas and the humid coastal belts of tropical countries. The total world area is eight million acres under coconut and the annual production of 14,000 million nuts.

Coefficient of correlation (Sta) When the relationship between two varieties x and y is linear, intensity of association is measured by what is called product moment correlation coefficient.

Coefficient of variation (Sta) It is a measure of variation independent of the unit of measurement and is therefore useful for comparison between different population as provided by the standard deviation expressed as a percentage of mean. This measure is known as the coefficient of variation and is given as:

$$CV = \frac{\theta}{u} \times 100$$

We say that the coefficient of variation is about 14 per cent.

Coffee (*Coffea* spp.) This perennial crop is grown for its seeds which when roasted, ground and soaked in boiling water yield a fragrant and mildly stimulating beverage.

Cold pack Method of canning–In this method fruits are cut into suitable sizes and filled in the can. Hot syrup is poured over these fruits in the can. Sometimes the fruits are slightly heated to make them soft and clean and thus more fruit material can be filled in the can.

Cold stores Co-operatives are also establishing cold stores to facilitate the storage and marketing of perishable commodities like fruits and vegetables, especially seed potatoes.

Cold test A germination test in which seeds are planted for a period in cool, moist unsterilised soil before transfer to a higher temperature, designed to approximate unfavourable planting conditions in order to seeding vigour.

Cold wave A sudden surge of cold air experienced in temperature attitudes usually caused by polar air being drawn equate forward behind the cold front of a passing depression. This is most prominent in the north continents where winter temperatures reach much lower levels than in the mainly southern hemisphere oceans.

Cold waves and frosts Cold wave or frost warnings are issued by radio broadcasts and telegrams, 24–48 hours in advance by the Indian Meteorological Department to enable farmers to adopt necessary protective measures. Usually the cold air flowing during the night collects at the bottom of a valley. Valleys in areas subject to frost should therefore be avoided for planting garden crops and orchards.

Cole crops This group of crops includes cabbage, cauliflower and knolkhol. They belong to the genus Brassica of the family Cruciferae. They are essentially cold weather crops and can withstand considerable frost; when grown under warm conditions, the produce becomes markedly pungent.

Cole crops sowing In north India, seeds may be sown from March to July in the hills. In the plains, the seed of early cauliflower may be sown from June to August and of late varieties from September to middle of October. Cabbage and Knolkhol are sown from August to October. In south India, the seeds may be sown from September to November.

Coleoptile (in Gramineae) The first leaf above the cotyledon which encloses the stem tip and other leaves; sheath.

Collective farming A system of farming in which the land is jointly owned or occupied by people who organise and run it according to a plan that the community has accepted or approved. The underlying concept of collective farming is the belief that there is a greater satisfaction and level of efficiency to be obtained from working for the community rather than for oneself.

Colour separations of seed processing Many large crop seeds such as peas and beans differ in colour between varieties; colour variations may also occur due to immaturity or disease. Electronic colour sorting machines can separate such a seed by differences in colour and also remove mud-balls and discoloured seed in the same operation.

Combine or thresher Harvested seed material can be run through a stationary combine with a second to remove awns and other seed appendages. A high cylinder speed and close concave setting are usually used to debeard or defuzz the seed. Cylinder speed and clearance setting must be such that excessive mechanical injury is avoided.

Combining ability, general The average performance of strain in a series of crosses.

Combining ability, specific The performance of specific combination, of strains in crosses. The relative ability of a bio-type to transmit desirable performance to its crosses.

Commercial centre A town or city that has a major function of buying and selling of goods and services. Some of the earliest commercial centres were the market town that traded in the agricultural produce of the locality. Present day commercial centres have banking, insurance and all other commercial facilities.

Commodities implications of world These will have a significant implication for the developed and the developing world.

Common market An agreement by a group of countries to form a single market within their group. The term is most commonly applied to the group of countries that form the European Economic Community (EEC) set up by the treaty of Rome on 1st January, 1958. The primary aims of the treaty are the eliminations of all obstacles to the free movement of goods, services, capital and labour between the member countries and the setting-up of common policies for external trade, agriculture and transport.

Companion crops These are such crop mixtures in which seeds of the crops intended to be mixed, are not mixed beforehand. They are sown in groups of separate lines/rows.

Compatibility of root stock and scion *See* Root Stock and Scion Relationship.

Compatibility of sprays Sometimes it is convenient and economical to spray two or more materials and then control more pests than one in a single operation. This can frequently be done but not all sprays are compatible *e.g.* lime sulphur is not a stable compound and must not be added to soaps and oils.

Complementary genes The non-allelic dominant genes that act together to produce a phenotype different from that produced by the homozygous recessives of the one or the other, both are called complementary genes. They give a phenotype ratio of F_2 as 9 : 7.

Competition of bacteria with plants The way in which that may be harmful at least temporarily, is by competition for available nutrients.

Complete fertilizers Manufactures supply such brands as to contain all three essential nutrients in one brand. These are termed as complete fertilizer brands.

Complete manure All the organic manures such as dung, compost and green manure meet all the elemental requirements of plants and thus are called complete manures.

Composite sample A seed sample composed by mixing different sub-samples taken from various parts of seed lot.

Composition of farm manure The term farm manure is applied to refuse from all animals of the farm, although as a general rule the bulk of the ordinary manure which ultimately finds its way back to the land is produced by cattle goats, horses and poultry.

Composition of papaya It is an important source of vitamins A and C and of calcium and basic ash. It ranks second one to mango as a source of the precursor of vitamin A. While this vitamin is generally associated with carotene, the yellow pigment in the papaya is not this but caricaxanthin.

Compost All the farm residues, animal castings and city garbage when decay, they form the compost manure. Sometimes the process of decay is performed by making compost pits and dumping the residues in these pits and leave them for decay in these pits.

Composting In vegetable production, well decomposed manure is mostly preferred especially if the soil is sandy. Such is the case with kitchen gardens, flower beds, lawns and green houses. Composting, which is largely a biological process, is often resorted to especially in the later cases, farm manuring being used in building up the compost pile. Because of its high organic content and influence of its nitrogen and micro-organisms, manure is especially valuable for such purposes.

Composted manure Another method of augmenting supplies is the preparation of compost from farmhouses and cattle-shed wastes of all types. These raw vegetable materials are rich in cellulose and other readily decomposable carbohydrates and have a carbon-nitrogen ratio of 40 or more to 1. The direct application of such undecomposited, low nitrogen organic matter as manure brings about a temporary deficiency of mineral nutrients (specially nitrates and ammonium compounds) in the soil by stimulating the growth of micro-organisms, which in turn compete with crop plants for available nitrogen-phosphorus and other elements. Hence, before using them as manure, it is necessary to compost or partially decompose them. Thus lowers the carbon-nitrogen ratio to about 10 or 12 to 1.

Compound fertilizers These fertilizers are multiple nutrient materials, supplying two or three plant nutrients simultaneously. When both nitrogen and phosphorus are deficient in a soil, a compound fertilizer like ammophos can be used. It contains 16 per cent N and 20 per cent P_2O_5. Its use does away with the necessity of purchasing two different fertilizers and mixing them in a correct proportion before use.

Condition of seeds health The extent of infection, diseases and insects. Results of a health examination are often required on the certificate.

Conditioner These materials prevent the fertilizer mixtures from forming clods. This facilities the easy spread of fertilizers in field. Materials such as tobacco stems, groundnut and husk of paddy form very good conditioners.

Condiments Such plants of which products are used in small quantities as culinary articles for cooking purposes.

Confounding of (Sta) The treatment combinations to be allotted to different blocks require to be grouped in such a manner as to ensure that the important comparisons in which we are interested are kept free from block differences. The important device of confounding certain relatively unimportant treatment contrasts with blocks plays a very useful role in many exploratory experiments in which a number of factors are required to be tried simultaneously in the absence of any previous knowledge regarding their interactions.

Consequence of waterlogging The principal results of water-logging conditions are the suppressions of changes of mesophytic vegetation in soil microflora and the growth of several forms of algae. The growth and activity of algae in submerged soil, some of which possessing efficient systems for fixing atmospheric nitrogen, are widely recognised in tropical rice field soils. The photoautotrophic habit of these organisms and their luxurious growth in aquatic environments explain their agricultural and ecological importance.

Conservation of soil and water Neglect of this idea may lead not only to shortage of water but also result in natural disasters like earthquake.

Consolidation of holdings and prevention of fragmentation The small size of the average holding has been one of the major obstacles to the improvement of agricultural production in India. The main reasons for diminution of the size of holdings has been the increasing pressure of population on land accompanied by the decay of village industries and the failure of employment in other occupations.

Constituent Included in.

Contact foliar herbicides These kill only the plant or portions of the plant actually contacted by the chemical which means net adequate coverage of the foliage is essential. Selectivity is dependent upon the differential wetting, differences in form of plant and upon placement of the spray. Some examples are decryl, potassium, cyanate, selective weed oil (carrot oil), sodium arsenate.

Contour farming One of the successful methods of controlling runoff is to carry out all cultivation operation and to sow the crops along contour. The plough furrows are all level in this case and hold the light rains till they are absorbed by the soil. Even in the event of heavy rains, the numerous stems of a dense stand of plants impede the flow of moving water, reduces velocity and lessens erosion. Farming on the contour thus reduces runoff, saves more moisture for crop production, reduces soil losses and increases crop yield.

Contour ploughing A technique of ploughing parallel to the contour of a hill slope rather than up and down the slope. It is widely used in semiarid areas to reduce soil erosion and gullying.

Contour strip cropping It is the growing of a soil exposing and erosion-permitting crop in strips of suitable widths across slopes on the level or contour, alternating with strips of soil protecting and erosion resisting crops.

Contour system At places where ground is too steep to allow irrigation of cultivation in straight lines, as in the hilly terrain, without undue erosion. Under such circumstances, the trees may be planted in lines following the contour of soil, with only a slight slope. Irrigation and cultivation are practised only along the tree rows, not across it.

Contour trenching It consists of a series of two feet by one foot trenches of suitable length excavated across the slope at suitable intervals. The excavated soil is deposited as a bund along the lower edge and tree seedling are plants alongside the trench just above it for a forestation purpose. Such trenches in a staggered fashion are more effective.

Contour terracing Series of properly spaced ridges or drainage channels are formed along the contours by the construction of suitable mounds of earth. They are called the 'level' terraces when constructed along contour and 'graded' if they are given water. The terraces are of four types, namely (*i*) channel terrace, (*ii*) board based ridged terrace, (*iii*) narrow based ridge terrace, and (*iv*) bench terrace.

Control and regulation of irrigation water Irrigation water must be controlled and regulated at all stages, ranging from the point of take off to the point of application. The velocity of water in-channels should be such that it may not cause erosion. The channels must be equipped with such regulation structures which may control the quality of water flow. Drop structures of various materials are installed to ease the slope of the channel-bed, for removal of undesirable gradient Diversion structures, with gates, are provided to divert and divide water flow and control its quantity.

Conversion factors The following factors will be found useful:

Distance	1 mile = 1760 yards = 5280 feet
	1 metre = 1,0936 yards = 39.37 inches
	1 yard = 0.1944 metre
	1 foot = 0.3048 metre
	1 inch = 2.54 cm.
Area	1 acre = 4840 sq. yards = 43560 sq. ft
	1 guntha = 1/40th acre = 1089 sq.ft = 33 × 33
	1 acre = 0.4047 hectare
	1 hectare = 2.471 acres
	1 sq. mile = 640 acres
Weight	1 long ton = 2240 pounds
	1 short ton = 2000 pounds
	1 metric ton = 1000 kilograms = 2204.6 pounds
	1 long ton =28 maunds (approx)
	1 maund = 80 pounds (approx)
	1 metric ton = 1.1023 short tons
	1 metric ton = 0.9842 long ton
	1 short ton = 0.9702 metric ton
	1 long ton = 1.0161 metric ton
	1 pound (avoir) = 453.6 grams = 0.4536 kg
	1 hundred weight (cwt.) = 112 pounds = 50.8 kg
	1 kilogram = 2.2046 pounds
Weigh per area	1 lb per acre = 1.12 kilograms per hectare
	1 cwt. per acre = 125.6 kilograms per hectare

Capacity	1 gallon (U.S.) = 3.7853 litres
	1 gar/on (Imp) = 4.5461 litres
	1 gallon (Imp.) = 0.1604 cft.
	1 litre = 0.2200 gallon (Br.)
	1 litre = 0.2642 gallon (U.S.)
	1 bushel = 56 lb (potatoes)
	1 bushel = 56 lb (barley)
	1 bushel = 60 lb (wheat)
	1 bushel = 42 lb (oats)
	1 gallon (Imp) of water weight of 10 lb = 1 cft. of water weights 62.3 lb.

Convolvulus arvensis Fam. Convolvulaceae. A perennial weed plant. For field *C. arvensis* (bindweed) 2, 4-D volatile ester alone to five lbs a.e/acre in 100 gallons of water at early bloom stage-control the seed.

Cooperation It is a method of organisation in which persons associate voluntarily and form themselves into a body for satisfying their specific needs.

Cooperative farming As the vast majority of cultivators in India consists of small farmers and landless labourers, cooperative joint farming on a voluntary basis is one of the method of reorganisation of Indian agriculture.

Copra The fleshy kemel of a coconut that has been dried. This is used in confectionery as desiccated coconut and is also crushed to produce coconut oil, which is used in the preparation of cosmetics and soap. The main exporting regions are Philippines, Indonesia, Sri Lanka and the islands of the SW Pacific.

Cordial Under the Agricultural Produce Act, 1937, the product is defined as juice when unsweetened, as cordial when it contains a prescribed quantity of sugar and is called squash when juice sacs have been included in cordial.

Cordon system of grape vine training After the vine has reached a height of two to four feet, the stem is bent and trained along the single wire trellis. On each arm that develops from the trunk, short fruiting spurs, each carrying two to four buds, are allowed to remain at the time of pruning. Replacing spurs are also provided close to the base of the fruitings spurs. *See* Trellis System.

Corms and rhizomes vegetables There are a variety of other crops in which underground parts are cooked and consumed. These chiefly include colocasia, pachyrrhizus, dioscoreas and elephant foot yam.

Corn The grain of seeds of any of the cereal grasses used as food, such as wheat, rye, barley, oats and maize. The word is used in certain regions or countries for the most important cereal crop of that area, *e.g.* corn is used in England for wheat, in Scotland and Ireland for oats and in the USA and Canada for maize or Indian corn.

Correlation (Sta) This is a simultaneous variation in two variables and is calculated by statistical methods. It often happens that changes in one variable are accompanied by changes in the other and that a definite relation exists between the two variables. In other words, there is a correlation between two variables. Analogous table for the frequency table for a single variable there is correlation table for the simultaneous distribution of two variables. The correlation table is a two-way classification similar to the one represented by contingency table. Some variables are positively correlated and some are negatively correlated.

Cotton A plant native to tropical and subtropical regions, several species of which are cultivated for the white fibre (cotton) that surrounds its seeds. The seeds and fibre are contained within the boll or seed pod, and it is shortly after this has split open that the cotton is harvested to avoid rain or wind damage. The fibre is spun, woven and dyed to produce cotton cloth. It is also blended with artificial fibres to make cloth. The seeds are separated from the cotton fibre on a machine called a gin; these are crushed to produce cotton seed oil used in margarines, cooling oil, soaps etc. and the residual pulp is used as an animal feed.

Cotyledons The parts of a germinating seed also called seed leaves.

Counts of staple In cotton, the number of 840 yards long threads present in one lb. of cotton ball determining the counts of cotton.

The finer the fibre, the numbers of counts are more. In thick fibres, the numbers of counts are less.

Cowdung It is universally recognised as being better than synthetic products for the retention of humus in soil and for maintaining its fertility. Its practical advantages over these products consists in the fact that it can be applied readily by farmers without the necessity of soil analysis and knowledge of the soil content.

Cover crop Any crop grown as to provide a cover to soil of an orchard. Such crops are useful in checking soil erosion and also checks the fertility of soil from deterioration.

Crab apple Have a number of species in cultivation.

Crag herbicide Same as Sesone. A herbicide.

Crawler tractor These are suitable for relatively high draft and low speed work. They can also be used on uneven ground or on soft-marshy land where the wheels would not work. These are heavier type tractors but their weight is more evenly distributed on a wider area; hence they do not compact the ground as much as the wheel types do.

Critical temperature A temperature that brings about a given response in a plant or animal. Freezing point 0°C, *e.g.* is a critical temperature for plants as it prevents photosynthesis. Critical temperatures especially winter minima are often critical in determining distributions of species.

Crop Plants grown for economic needs such as for food, clothes, oil fodder, to increase the fertility of soil and to earn a living, are called crops. These are divided in according to season they grown in such as winter crops, summer crops, zaid (in between) crops.

Crop mixtures Crop mixtures have the added advantage of acting as insurance against total crop failure in an unfavourable season. Much additional information is, however, needed to evolve cropping and manurial system for both irrigated and unirrigated conditions to control erosion and improve or maintain soil fertility most effectively.

Crops for green manuring The crops most commonly used as green manure in India are Sannhemp (*Cratolaria juncea*), dhaincha (*Sesbania aculeata*), Cluster beans (*Cyamopsis tetragonoloba*), Senji

(*Melilotus barviflora*), Cowpea (*Vigna catjang*), horse gram (*Dolichos biflorus*), Pillipesara (*Phaseolus trilobus*), berseem (*Egyptian clover*). Lentil (*Lens esculenta*) is recommended in Kashmir for green manuring paddy. In consequence of the varying influences, soils of different textural classes are considered suitable for different crops. Thus, paddy, jute and dhaincha (*Sesbania aculeata*) are well suited to clay or clay loams, whereas wheat, gram and cotton grown best on loams and sandy loams. Similarly groundnut, sorghums (particularly bulrush millet) and Potatoes and turmeric fare well on predominantly sandy. soils.

Crops for hay making Grass as obtained from reserved grasslands is the main material made into hay. Principal grasses used for this purpose indifferent parts of the country are *Cenchrus ciliaris* (anjan), *Dichanthium annulatum, Dichanthium caricosum, Heteropgon contorus* (spear grass), *Iseilema laxum, Iseilema wightii, Schima nervosum* and *Themeda* spp. As a rule, thin stemmed and leafy grasses make better hay than the coarser ones.

Crops tolerant to alkali The growing of alkali-tolerant craps and plants is yet another method advocated sometimes. Sugar-beet, rice, lucerne, patsann (*Hibiscus cannabinus*), wild indicio and babul (*Acacia arabica*) are plants of this type.

Crop residues for animals use The dry straw of cultivated cereals like jowar or cholam (sorghum), bajra or cumbu (*Pennisetum typhoideum*), paddy, maize, wheat and ragi (*Eleusine coracana*) forms the most important source of cattle feed. Stems and plant residues of leguminous crops like ground nuts, gram, other pulses, guar (cluster beans) and cowpeas are also commonly fed. Except jowar straw (commonly called kadbi) which provides a maintenance diet, other cereal straw are generally coarse and deficient in protein. Residues of leguminous crops, however, are most nourishing.

Crops' response to farm manure The manure is available on the farm in a limited quantity and as mineral fertilizers give good results on grass and legumes, it is often considered judicious to reserve most of the manure for other crops in the rotation. Crops like maize, cabbage, tobacco, cotton or potatoes respond better to the farm manure.

Crop response to liming The leguminous crops respond most markedly to lime application, though there are differences among the various legumes. Of the non-legumes, asparagus, spinach, sugar beets and garden beets, cabbage and cauliflower are especially influenced by lime even if the soil is not strongly acidic.

Crops suitable for green manure All leguminous plants and among non-legumes rye, oats, barley, millet, buckwheat, wheat, mustard, weeds and the like are suitable.

Crop rotation The importance of crop rotation in the general maintenance of soil fertility is well known. Planting the same crop year after year depletes the plant nutrients in the soil in an unbalanced manner. Furthermore, the uninterrupted growing of row crops needing frequent clean cultivation also causes increased soil loss by encouraging sheet and gully erosion. A proper rotation of crops will control erosion, lessen soil loss and preserve productivity. A good rotation should include a cultivated row crop, densely planted small grains and a spreading legume or a legume and grass mixture. Naturally, the actual choice of rotation crops in every tract will be based on land use capabilities taking into consideration the climate, economic conditions soil type, slope, character and degree of erosion, and other physical and chemical properties of soils. In many places, the growing of crop mixtures, which usually include pulses, takes the place of or is practised besides rotation involving recurrent succession of different crops.

Crop rotation benefits It is very important in maintaining soil productivity–(1) They are practically as effective as farm manure in case of certain crops, (2) The benefits of crop rotation do not impair those received from manure and commercial fertilizers, and (3) Compared with manure and commercial fertilizers, the influence of crop rotation containing a legume, farm manure are more beneficial.

Crop sowing and rotation The sowing of crops in a given period in any land in such a sequence that fertility of the soil does not deteriorate and the production be high, meets the successful requirement of crop sowing.

Crop weather calendars The yield of a crop is determined by many factors such as weather conditions, variety grown, incidence of insect pests and diseases, fertility status of the soil, manuring, irrigation and other cultural practices. As such, a proper knowledge of the relation between crops and the weather risks they face at successive stages of plant growth, is necessary for profitable farming and is done by regularly maintaining crop weather calendars.

Crop-weed competition In non-irrigated areas the competition between weeds and crops is largely for water. A saving of 300 to 500 tons of water in an acre of soil, six feet deep, is possible by keeping it free of weeds.

Cross inoculation of legumes This is done in case of one strain of bacteria irrespective of how good results are shown on a certain legume. The organisms are interchangeable and thus their species on the legume may prove suitable and be effective by the available strain of bacteria.

Cross ploughing To plough the field in two directions.

Crown In pineapple, the crown or top of the fruit is sometimes planted and under favourable conditions will grow well, although in ordinarily takes at least two years to produce ripe fruits.

Crown grafting Also called rind or bark grafting. The scions are cut diagonal as for splice grafting and are inserted between bark and wood. Also *See* bark grafting.

Crystallised fruit The method of preparation of crystallised fruits is the same as for preserves. After the preserve is ready, the syrup is drained out by placing the preserve on sieves, and the pieces of fruit rolled into powdered sugar and dried in shade. Whole small fruits or pieces of larger ones are slowly impregnated with sugar. Any floating fruit should be kept under the syrup.

Cube is an important source of insecticide. It is obtained from two species of leguminous genus Lonchocarpus, is also grown in Malaya. Also *See* Botenone.

Cucumber tree *Averrhoa bilipulin*, Fam. Oxaliaceae.

Cucurbitaceous crops The crops falling under this group, *viz.*, cucumbers, melons, pumpkins, squashes and gourds, belong to the family Cucurbitaceae. They are hot weather crops and are mostly trailers with the exception of bush varieties of squash. They are insect pollinated and the flowers are usually unisexually both the kinds occurring ordinarily on the same plant with the exception of parwal (pointed gourd) where they are on different plants. Pointed gourd and small gourd (tondli) are perennial and the rest are annuals. The ripe fruits of musk melon and water melon are taken as dessert whereas the unripe, tender fruits of all these crops are cooked as vegetable. They are a good source of vitamins.

Cull *See* Sorting.

Cultivar Any agricultural of horticultural 'variety'. The term is derived from the words cultivated variety.

Cultivators Cultivators are either 'spring tyned' or 'rigid tyned' cultivating devices. The formers have a fine pulverising effect while the latter are good for deep cultivation on light lands. The average output of an eight-foot wide cultivator with 11 tynes is about 1.5 acres per hour. Heavy cultivators fitted with side shares, going two to three inches deep even in hard soils, can be used for stubble clearance.

Cultural or cropping methods Weeds under many conditions are better competitors than crop plants for light, water and mineral nutrients. However, farming practices are capable of affecting the degree to which crop plants can successfully compete with weeds. Strong germinating seed will give the crops a vigorous close stand with a head start on the weeds may make it difficult for them to get under way. Varieties which are well adapted to a growing region will obviously compete better than poorly adapted varieties. The plant breeder has to evolve quick growing varieties of crop plants with a large leaf area and good branching, and the agronomist has to work out the proper seed rate, depth, time and method of sowing, irrigation and fertiliser requirements, placement of fertilizer, proper systems of rotation of crops and plant protection measures.

Cultural practices In raising a crop, certain work has to be done as to find out requirements of the plant and this type of method of growing crop.

Cumulose This is of recent origin; made up largely of partly composed remains of plants.

Curing and fermentation tobacco Tobacco leaves are cured after harvesting in order to impart the required texture, aroma and colour to the final product. Virginia tobacco is flue-cured in bams, in which hot air is circulated through iron pipes known as flues. It takes 100 to 120 hours to complete the curing process. *Bidi* tobacco is usually ground-cured leaving the cut plants on the ground till the laminas of majority of leaves become dry. The plants are then collected and laminae served off the mid-ribs. The laminae are taken to store and the mid-ribs are allowed to dry in the threshing yard. Lamina pieces adhering to the mid-ribs are collected as green. In Nipani, the produce is stored and sold as anghad which consists of lamina, geran and sand leaves while in Gujarat these fractions are stored and sold separately. In cigar and cheroot tobaccos, the cut plants after wilting for a day are gathered in the morning in circular heaps which are opened up in the evening and remade next morning repeating the process of 10 days for air-curing and two days for rack-curing. The air-curing is done in *pandals* while rack-curing is done in the open till the mid-ribs dry out. The cut plants are then brought to condition, heaped and the leaves stripped and made into hands of 75 to 100 leaves. Chewing tobacco in Madras is rack-cured, air-cured, pit-cured or smoke-cured (Jaffna method) in kilns. Hookah and chewing tobaccos are ground-cured in Bihar, U.P., West Bengal and Gujarat and pit-cured in Punjab.

The cured leaves, except of bidi, are the bulked into convenient sized bulks for fermentation which improves the quality of the leaf and aroma. The bulks are broken up and the material shaken occasionally to avoid over-fermentation and decay. The leaf is then graded in accordance with trade requirements.

Curcuma longa Turmeric suffers from the attack of a serious leaf spot, the casual organism being *Taphrina maculans*.

Currants The raisins made from two varieties of seedless small grapes grown mainly in Greece are known as currants. These are different from Ribes-currants.

Current net worth The net work is the value of assets minus the liabilities of the entire farm business.

Custard apple *Annona squamosa*, Fam. Annonaceae. Custard apple of Sitaphal is a small tree seldom more than four and a half metres high and is both cultivated as well as found semi-wild growing throughout peninsular India. A dry and hot climate suits it best. It flourishes on lighter types of soils and grows even on slopes of hills, but cannot stand cold or frost. Propagated by seeds as well as by grafting. There is considerable variation in fruit when seed-propagated. Bullock's heart (Ramphal) provides a suitable rootstock for grafting by inarching or budding. Grafted plants give fruits earlier (about two years) than seeding which takes about three to four years. The plant flowers from April to May and fruits are ready between August and November. The yield is almost 25 to 30 kg per tree. Fruits are mostly consumed locally but they can be transported over a long distance, if they are picked before they are fully ripe and packed in single layers in well-ventilated wooden boxes with a soft padding material.

Cyclone An area of low atmospheric pressure in the tropics, 80–400 km in diameter. It is a storm system of great intensity winds circulating at speeds of 120–280 km per hour around the 'eye' (centre) of the cyclone and accompanied by torrential rain. Winds rotate in an anticlockwise direction in the Northern hemisphere and in a clockwise direction in the Southern hemisphere. Cyclone can cause considerable damage to agricultural related commodities and buildings, and are a serious hazard to shipping.

Cyclonic rainfall Precipitation along the frontal surface of depression in mid and high latitude. It is caused by moving over the cold heavens air of another or by air rising through horizontal conversions in an area of low pressure. Drizzling rain falls in a broad belt followed by more concentrated squally rain as the cold front passes.

Cynodon plectostachyum Star grass. A good fodder grass.

Cyperus rotundus Fam. Cyperaceae. A perennial weed. Can be controlled by 2,4-D volatile ester 2 lbs a.e/100 gallons; spray as many times as necessary when it is 4–6 inches tall. Amino triazole 2 lbs/acre plus wetting agent 6 fluid ozs.; spray at active growth stage; repeat if necessary. Eptam 5–10 lbs/acre in 80 gallons of water, apply by soil incorporation.

D

Dalapon 2-2 dichloropropionic acid (Na-salt), a herbicide. It is a systematic herbicide. The introduction of dalapon as a systematic killer of annual and perennial grasses has put in important new tool into the hands of farmers. Dalapon has proved to be very resistant to growth in plants.

Dam A structure built across a river. A dam may be constructed to provide water for irrigation, industrial use and human consumption; to provide a head of water for hydroelectric power (HEP) generation; to divert the water flow for navigation; or to control the water flow to prevent flooding. Frequently a dam fulfils more than' one purpose and may also be used for recreational purposes.

Damage by frost The severity of the damage caused by frost depends not only on actual temperature and the duration of frost, but also on condition of the plants. Young plants are more susceptible to damage than old ones and new growth is more likely to be injured than that which has had time to mature.

Damping off Caused by fungus *Pythium aphanidermatum* and *P. myrotylum*.

Dark blocks Such areas where ground water falls short of the local demand.

Date (also called date palm) *Phoenix dactylifera* and *P. sylvestris*, Fam. Palmaceae. In this tree, pollination is inhibited by direct sunlight but takes place at night or on cloudy days. Some of the imported varieties of date palm have been cultivated successfully in the Punjab. It thrives in light as well as heavy soils provided depth is adequate (2.5 to 3 metres) and drainage good. The plant is unaffected by frost and requires intense heat in summer for the development and ripening of the fruit.

Dead seed Seeds incapable of germinating.

Debearder Oats, barley, some vegetable and flower seeds and many grass seeds have awns or appendages that make them difficult to process and plant Hairy appendages on certain grass seed make them rubbed together or are rubbed by mechanical devices, until all appendages are removed. This rubbing action will also break up clusters or dumbbells found in many small grass seeds. Threshers, hammer mills and debearders are used to remove troublesome seed appendages.

Decay Some of the intermediate complex products of vegetable matter decay are resins, waxes, oils, fats and lignins, etc. These are of organic decays.

Decay of green manure It has some advantages such as (1) the general purpose organisms of the soil become active along with the decay bacteria, (2) due to much increase of microbial tissue and demand for, and the turnover of, both carbon and nitrogen are exceedingly great, and (3) on the slackening of the vigorous decay, simple end products other than carbon dioxide begin to appear along with humic remains of the green manure.

Deciduous Plant shedding their leaves.

Decoction A liquid made after boiling a plant material for a long time.

Dichogamy *See* Plant Science.

Decomposer An organism that feeds upon dead organic matter breaking them down into simpler substances. Decomposers recycle nutrients making them available to producer organisms. Bacteria and fungi belong to the decomposer group of an ecosystem.

Decomposition level A chemical reaction in which a compound is broken down into compounds with simpler molecule and forms.

Decomposition of organic matter There must be adequate moisture in the soil for the proper decomposition of organic matter. The farmyard manure can, therefore, be applied to all crops grown in the rainy season or under irrigation.

Deep fry Frying in a deep-medium.

Deep placement Fertilizer is placed at about 3 to 4 inches deep. It has shown better utilization of manures at this depth.

Deficiency Lacking in something.

Deficiency disease of guava (*Psidium* guava) Caused by zinc deficiency. It causes severe reduction in leaf size; interveinal leaf chlorosis; suppression of growth; die-back of leaders and extreme reduction in yield of fruit; fruit of poor quality. Foliar spray with commercial sulphate. Soil application of zinc sulphate @ 4–12 ozs. per plant should be applied to control.

Defined Well marked.

Definite True to type.

Degree of alarm in use of herbicides It is not necessary to become unduly alarmistic in the use of herbicides that are imported into our country because all these have been tested and found safe for use in crops.

Dehydration When water becomes scanty in a system.

Delayed foliation *See* Rest period in temperate fruits.

Dietic value of fruits This has been now proved and the public is also realizing now that inclusion of fresh fruits and vegetables should be a matter of necessity in the regular diet. The demand for fruit is increasing among the educated classes and will continue to increase as knowledge of its value spreads.

Density of population of microorganisms The soil organisms vary in numbers from a few per acre to many millions per gram of soil. The density of population is determined by food supply, moisture, temperature, physical condition and reaction of soil. In more or less neutral soils, bacteria dominate the microfloral life. If the soil is acid and rich in organic matter, fungi predominate. Algae bound on or near the surface of soil in constantly moist, and shady localities.

Depletion of nutrients In gas form, only nitrogen is lost from soil. As a result of various chemical reactions in the soil, nitrates are converted in elemental nitrogen. These are in the form of gases and whenever such time comes the escape in the form of gas.

Depletion of nutrients by crops The biggest depletion from the soil of the nutrients is done by crops. The nutrients are taken by crops in the following amounts:

Sl.No.	Crops	Production per Hectare in Kgs	Per Hectare Depletion of Elements from Soil in Kgs		
			Nitrogen	Phosphorus	Potassium
1.	Paddy	2500	71.5	21	110
2.	Wheat	2500	63	31	75
3.	Jwar (Sorghum)	1500	70	25	115
4.	Bajra	1250	40	15	140
5.	Linseed	1250	29	18-	38
6.	Maize	2500	105	45	100
7.	Cotton	625	45-	23	68
8.	Jute	1875	85	25	50
9.	Groundnut	2225	90	20	50
10.	Barley	2500	55	30	125
11.	Sugarcane	125 tonnes	170	100	495

Deposition To settle at the bottom or any part of the body.

Depth of ploughing Best inversion is secured with mould-board ploughs when the depth is less than 2/3rd of the width of cut of each furrow bottom. A working-depth of less-than 1/4th of the bottom width would result in poor quality of work. So, a 12 inch plough can work well for depths varying from three to eight inches. A digger type plough cuts a deeper furrow almost equal to its width, but cannot deal so successfully, with heavy trash. The shape of the share particularly its vertical section affects penetration.

Derris Practically derris is in Malaya where it is grown as a profitable crop.

Desirability of green manures An ideal green manuring crop should have three characteristics–(1) rapid gowth, (2) abundant and succulent tops, and (3) the ability to grow well in poor soils also.

Desiccating winds Hot dry winds causing damage to plants and fruit trees.

Despatch of seed samples Samples sent by post should be securely tied. Sealing is desirable. An alkathene bag holding a sample for moisture test should be included with the major sample if this test is desired. Cotton or jute bags, heavy card-board or similar containers are desirable. It has to be assured that a container is clean before putting the sample for test in it.

Detached method of grafting In this case, the scion branch is separated before it is completely joined with the stock. Various types of methods such as whip or tongue grafting, wedge grafting, crown grafting, side grafting, saddle grafting, top grafting, root grafting and bridge grafting; all come in this category.

Deterrent in use of compost Social prejudice against the use of this valuable compost has disappeared and town composting is being rapidly adopted in other localities all over India. With suitable modifications, such as the provisions of trench latrines, it can be taken up in village also.

Development of irrigation systems The future development of irrigation aims, at impounding the surplus river-flow during the monsoon for use during the dry weather. Tanks which form an important source of irrigation are mostly rainfed. They depend for their replenishment on the surrounding drainage area and watersheds. As runoff waters from the drainage area carry large quantities of sediment tank sites are usually selected in watersheds that are protected by vegetation. Where this is not possible, suitable soil conservation practices are adopted to remove as much sediment from runoff waters as possible.

Dewberries *Rubus allegheniensis*, Fam. Rosaceae.

Dew point The temperature of air at which it becomes saturated with water vapour; below this point water vapour starts to condense to form water droplets. It can be determined by using a hygrometer with tables used on dry and wet-bulb temperatures or by using a special dew point hygrometer with, a polished material surface that can be cooled until a film of moisture appears at a temperature that is recorded.

Diagnosing fertilizer needs of soils There are four methods of determining the fertilizer requirements of soil–(*i*) field experiments, (*ii*) pot tests, (*iii*) biological tests, and (*iv*) chemical tests.

Diagonal cut In grafting, the two plants parts are cut diagonally as to completely fit into each other.

Diameter Any line passing through the centre of a circle.

Diathane Chemically it is the disodium ethylene bisdithio carbamate.

Dibbling of manure When the plants are at a distance from each other, a hole is made with the help of stick near the plant and manure is put in this hole.

Dichanthium annulatum A good fodder grass.

***Dichanthium cenchrus*, elyonurus cover** The characteristic perennial grass species are *Dichanthium annulatum*, *Cenchrus ciliaris* (Anjan), *Cenchrus setigerus*, *Elyonurus hirsutus* (Sewan), *Cynodon dactylon* (Dhub), *Eleusine flagellifera* (Chhimber) and *Soorobolus pallidus*.

Dicotyledon *See* Plant Science.

Dioecious *See* Plant Science.

Die back A plant disease caused by several fungi in many types of plants.

Digera arvensis Farn. Amar-anthaceae, A serious weed of summer season crops. *See Celosia argentia* for control.

Digging sapling from nursery It is convenient to dig a trench along each side of the row, with a spade or *Phawara*. This should be about four inches from the row and about a foot deep. The plants may then be easily removed with a ball of earth around the roots with the help of a *khurpi* (hand hoe), being an excellent tool for this purpose. The size and shape of the ball will depend on the size of the plant, the nature of its root system and the distance it has to be removed. The larger the plant is, larger is the ball of earth should be. If, however, the plant is to be shipped some distance, it is wise to decrease the size of the ball and prune the top more heavily. Shape of the ball will depend on whether the root is tap-root or several spreading roots (fibrous). For an ordinary tree with roots two or three years old, a ball of earth 15–20 cms in diameter and 15–25 cms deep is sufficient.

Diploid *See* Plant Science.

Dip method Disinfection of seed by use of chemicals or heat is a well known method of avoiding diseases. Chemicals also can be applied by the dip method or dipping the seed in the clinical solution.

Diquate-I 1,1-ethylene=2, dipyridilium dibromide.

Direct influence of humus on plants Humus to some extent can be considered to serve as a direct source of plant nutrients, providing not only nitrogen, sulphur, but also carbon, oxygen and hydrogen elements, usually considered as coming from air and water.

Disc coulter This disc coulter is suitable for all types of ploughs; it cuts cleaner and has a lighter draft than the knife coulter. The cut away disc and the wavy edged disc coulter are specially useful for working in a surface growth of creepers and tangled green manure crop.

Disc plough A disc plough cuts a radiused furrow whose width varies with the operating depth. The optimum depth is that which gives a width of cut equal to $1/3^{rd}$ the diameter of the disc. Unlike a mould-board plough-penetration is influenced also by the implement weights. So there is always some provision made for adding weight to secure penetration under hard conditions. The shape of the discs particularly the concavity affects inversion and pulverisation. A mould-board plough is more efficient than a disc plough if considerable amount of surface vegetation has to be buried: But disc plough is not susceptible to damage from stones, roots and stumps

Discrete variation (Sta) A distribution in which values of the frequency only exist at certain separate points of the abscissa scale. It may be represented by a frequency polygon, but should not be represented by a continuous curve.

Distribution The spreading of anything in an area, country or internationally.

Distribution of sample means (Sta) The behaviour of the means derived from random samples can be test studied with the help of an actual example of sampling. It is recommended that the worker should obtain a number of samples (100 or more) of different sizes with the help of tables of random numbers from the population of say 400 samples.

Diuron 3-(3, 4-dichlorophenyl)1, 1-dimethylurea, a herbicide.

Division To propagate plants by some part of the mother plant.

Doder (*Cuscuta* spp.) Has a long slender yellow stem which entwines itself tightly around the host, penetrating the bark at intervals with Haustoria, which help suck the sap.

Dodder mill Same as the roll mill. Since the separation of the rough dodder seed from smooth lucerne seed is a common use for this machine, hence the name Dodder mill.

Dominance *See* Plant Science.

Dormancy A condition where the seed will not germinate normally without a period of rest after ripening.

Doses of NPK Recommended doses of nitrogen, phosphate and potash fertilizers for specific crops should be applied to get good yields.

Dot map A map showing distribution by means of dots, each dot of a uniform size representing a specific value. The dot is the simplest form of symbol used on maps and the result is very effective in the value represented by each dot is appropriate. Examples include maps showing the distribution of farm animals such as cattle and sheep.

Double cross A cross between two F_1 hybrids.

Double purpose-crops Very often berseem, senji and lucerne (*Medicago sativa*), and sometimes sannhemp are sown partly for fodder and partly for green manuring. In the case of annual crops of senji and berseem, one or more cuttings are taken for use as green fodder. Lucerne, which is allowed to grow for two or three years, is cut seven to eight times for the same purpose. In sannhemp, the tops are fed to cattle. In all these instances the residues (roots and stumps) are incorporated into the soil. These crop residues contain considerable amounts of nitrogen, phosphorus, potassium and other mineral nutrients besides the organic matter. In the case of orchards, the annual green manure-cum-forage crop should be grown at such a time as to interface the least with tree growth and fruit development.

Double working In horticultural trees if the fruit to be grown is not compatible with the root stock best suited to the locality, it is sometimes possible to make first graft with a variety which is compatible with both and the graft the desired one variety on this.

Downy myrtle *Hodomyrtus tomentosa*, Fam. Myrtaceae.

Dowpon Same as dalapon.

Drain A channel or pipe to carry away excess water or sewage from the land. In the Fens of E. England, the term is used for the wide canals used to carry water to the sea or a river.

Draft of plough Draft of a mould-board plough varies from 2 to 3 lb. per square inch of cross sectional area of furrow slice in light soils to as much as 29 lb. per square inch of extremely heavy sticky clay soils. Under summer conditions in kans-infested lands, the unit craft is as high as 30 lb.

For a 40 H.P. tractor, a three bottomed plough having 12-inch wide mould-boards and weighing about 1,000 lb. is usually suitable. It will plough up to eight inches deep and cover about five acre a day.

Drainage of agricultural land Irrigation and drainage cannot be separated from one another. They are like two faces of a coin. While irrigation is meant to arrange adequate moisture in the soil for satisfactory growth of plant, drains are important for avoiding excess moisture in the root-zone. For improved cultivation, both irrigation and drainage are equally important, so that the plant has just the requisite supply of moisture nothing more nothing less. To avoid excess of moisture, care must be taken to dispose of any collection of water either above or below ground surface. Drains that deal with above surface water are called surface drains while those that deal with underground water are called seepage drains.

Dried bananas Green bananas or plantains are also dried and ground, forming what is variously called banana flour, plantain meal, pisang starch and banana meal.

Dried blood or blood meal It contains 10 to 12 per cent highly available nitrogen and 1 or 2 per cent phosphoric acid. It is a very quick acting manure and effective on all crops and all types of soils. It should be used the same saline way as oil cakes.

Drilling This method is used while giving the manure between rows with the help of machines.

Drop structures These are installed to ease the slope of the channel bed for removal of undesirable gradient.

Drought A period of dry weather, especially one lasting long enough for vegetation to begin to wither. Generally an absolute drought is defined as a period of at least 15 consecutive days, none of which receives more than 0.25 mm of rainfall, a partial drought is a period of 29 or more consecutive days with an average daily rainfall of less than 0.25 mm. In the USA, a drought is defined as a period of 21 days or more when the rainfall is 30 per cent or less of the average for the time and place.

Dry blood It has 8–14 per cent nitrogen, 0.9 per cent phosphorus and 0.7 per cent potash.

Dry-bulb temperature The shade temperature as measured on a simple mercury thermometer (dry-bulb thermometer). Simultaneously readings of this and a wet-bulb thermometer are used to calculate the relative humidity. *See* wet-bulb temperature.

Dryers Devices used in chemical processes to remove a liquid from a solid by evaporation. Drying equipment is classified by the method of transferring heat to a wet solid. This can be by direct contact between hot gases and the solid (direct dryers), heat transfer by conduction through a retaining-metallic wall (indirect dryers), or infrared rays (infrared dryers).

Dry farming The farming methods that are practised in semiarid regions to catch and store the limited moisture available so that crops can be grown. The main method used is to grow a crop every other year, breaking the surface to prevent it cracking at intervals during the fallow year to conserve the moisture for the crop of the year that follows.

In several parts of India, rainfall is very low, uncertain and highly variable, and ordinary farming is more or less a gamble. Two types of agriculture are possible in such areas of low and precarious rainfall, namely, of crop production and animal husbandry including management of grazing lands. In India with a high population, crop production or arable farming has to be restored to. The system consist of:

(1) in using special crop varieties (draught resistant of crops which need less moisture by their early crop period, (2) in reducing the rate of seeding to correspond to the limited moisture available, and (3) in employing tillage practices that conserve an economic moisture.

Dry farming and nature of crop Light shallow or partially eroded soils should be used for growing summer season crops of bajra and tur mixture of groundnut instead of rabi jowar. Such areas may be utilised to grow a green manure crop which should be harvested. and buried in the adjoining deep soils.

Drying It is the simplest method of preserving fruits and vegetables and may be of some advantage under some situations. It is desired by drying them in sun. It is commonly used in such fruits as grapes, figs, plums, peaches, apricots and some root-vegetables.

Dry spell In the UK a period of 15 or more consecutive days, in which no day receives more than 1 mm of rainfall. The term, however, is often used generally to describe periods that have a withering effect on plant life.

Dust herbicides Few herbicides are applied as dusts because of the drift hazard, though insecticides and fungicides are very often made and applied in the form of dusts.

Dusting machines Used to dust insecticides and chemicals etc. on the affected trees, crops or fields.

Dusters and sprayers As most vegetable crops are usually affected by a variety of fungus and virus diseases and a number of insect pests, it is better to equip oneself with the necessary sprayers that are available in the market and the choice must depend on the work to be done. Hand dusting and spraying machines are useful for kitchen gardens, while knapsack and rotary dusters and knapsack pressure sprayers are necessary for vegetable cultivation on an extended scale. Power dusters and sprayers are particularly useful in case of very large areas.

Dwarfing It is commonly practised in fruit trees and may be of some advantage under some situation. It is desired to have a large variety of fruits in a small area, the advantage of dwarf plant is obvious.

Dwarfing effect This effect is not only on the tree as a whole but partially on the part pruned. If two branches of equal size are taken out of which one is pruned more heavily than the other, the light pruned branch will become the larger, both in length and in diameter.

Dwarflike To look smaller in size.

Dwarf ornamental hedges Such plants as *Clerodendron interne, Dedoniaea viscosa, Duranta plumerii, Justicia gendarussa, Lawsonia alba* and *Myrtus communis* are some of the plants that come in this category.

Dwarf protective or armed hedges Plants such as *Acacia modesta, Citrus vulgaris, Clerodendron phlomoides, Duranta spinosa, Inga dulcis* are in this category.

E

Earthing This is an operation in which soil is heaved upto certain level on the stem as in sugarcane and potato and other crops.

Earthquake A series of shocks that result from sudden earth movement *e.g.* along a fault which can be recognised by the passage of earthquake waves (seismic). The waves spread out from the seismic focus of the earthquake and call cause widespread destruction.

Echinochloa crusgalli Fam. Gramineae. An annual weed. In rice, Stam F. 34 3–4 lbs./acre at 1–3 leaf stage. Many such annual grasses thrive in rainy season summer. Far annual grasses in cotton treat with diurion pre 1 lb./acre. In maize and sugarcane, simazine pre 2–4 lbs./acre is effective against such annual grasses.

Ecology *See* Plant Science.

Ecological regulation Natural regulation (balance of nature). A mechanism by which populations within a community are controlled about an optimum level. Limiting factors within the ecosystem such as weather, food supplies or predation, growth by suppressing a species' birth rate or increase its death rate. Such regulation is frequently density dependent *e.g.* it operates progressively more strongly as the population density grows.

Ecological regulation is the principle underlying (biological as opposed to chemical) control of pests.

Economic utilisation of farm manure Fineness of division and evenness of distribution are two important points to be kept in mind. With these, the efficiency of manure may be enhanced considerably. Also, the rate of application is important. Since the supply of manure is limited, it is better to decrease the amounts at each spreading and cover a greater acreage.

Edge Sometimes in a part or a garden, a differentiated row of small shrubs is planted as to add to the beauty of the place.

Edging Planting of small plants or bricks on the edges or boundaries is of beds of flower-plants is called edging. Far this, the common ornamental leafed plants such as *Alternenthra* spp., *Anthericum liliastrum, Aspistra* sp., *Caladium humboldii, Cineria maritima, Iresine* sp., *Justicia gendurussa, Phalaris arundinaceae, Pyrethrumaureum, Santolina chamaecy parissus, Coleus* and *Echeveria* are the common plants used for edging.

Effect The resultant outcome of any action.

Effect of boon on plants Keeps calcium in soluble state. Controls potassium/calcium ratio. Enhances absorption and activity of calcium in plants. Helps absorption of nitrogen. Mare auxiliary roots grow from the vascular part of plants and from these roots the bacteria in nodules get their food.

Effect of calcium on plants This is of use in the making of cells and in its presence plants become hardy. Root-growth starts sooner. Checks the ill and poisonous effects of organic acids found in plants. Improves the in-take of nitrogen in plants and in such condition, plants absorb more nitrogen in its presence. Encourages the function of flower and fruit bud formation. Chances of lodging of crops become remote.

Effect of copper on plants Controls the respiration activity. Enhances use of iron. Encourages the processes of oxidation and reduction.

Effect of iron on plants In absence of this element, plants suffer from chlorosis. It also helps in absorption of other elements. Formation of protein in the chloroplasts, presence of iron is necessary. It encourages the oxidation and reducing enzymes.

Effect of magnesium on plants Being a major constituent of chlorophyll, it is necessary for all green parts of plants. It acts as a carrier of phosphorus in plants and because of this activity is more helpful when this takes place in the oilseeds plants. It encourages translocation of starch in plants. Controls the absorption of other elements in the plants. It is helpful in the manufacture of oils and fats in plants.

Effect of manganese on plants Helps in making of chlorophyll. Acts as a catalyst in the oxidation and reduction processes in plants.

Effect of molybdenum in plants Helps in establishment of nitrogen. Encourages the process of oxidation and reduction.

Effect of nitrogen on plants Induces growth of the vegetative parts. Produces deep green colour in leaves. Grains become plump big in size. Succulence is brought about in parts of plants by which they develop lodging tendency. Excess of this element in soil delays the maturity of crops. Ratio between grain and husk is reduced. Succulent plants are more damaged by insect and disease pests. Protein and water contents of plants increases.

Effect of phosphorus on plants Growth enhances in plants, stems and pedicles become strong. Has a good effect on root-growth. Weight of grain increases. Induces resistance in plants against the attack of diseases and insect-pests. Protein content of plant increases. Nodules present in the leguminous plants increase both in number and size and thus enhances the nitrogen fixing capacity of such plants. Crop matures early. Flowering and seed formation are encouraged.

Effect of potash on plants Photosynthesis takes place more rapidly in plants. Food material starts going to other parts of plant from leaves more easily. Plants become healthy and green plants become strong. Grains develop attractive colour effect of nitrogen and phosphorus is more activated. Growth hastens in plants and their capacity to resist attack of insect pests and diseases increase. Crops do not lodge and grains and seeds become more plump. Leguminous and other plants develop more resistance against cold. Plants produce more protein. Less losses incur during transpiration of water from the cellular water. Enzyme activity in plant is enhanced.

Effect of sulphur on plants Though this is not a constituent of chlorophyll, still it has been found helpful in chlorophyll making. Plants attain much vegetative growth. If one of the creative part of manufacture oils and proteins in plants is sulphur, it encourages root-growth, seed formation in plants and nodule formation in plants of leguminous crops.

Effect of zinc on plants This is essential element used in relatively small amounts for plant growth. Its deficiency in citrus crop results in many kinds of diseases, retarded growth. Its ionic form found in the soil is Zn^{++}.

Efficacy of the management of fertility Any fertility system is highly flexible yet simple. All of the phases are easily adjusted and may or may not be applied, depending on conditions. This is true especially in respect to lime, to potash and to trace elements. In any case, it will generally pay to add a phosphatic fertilizer to a soil while nitrogen and organic matter almost always need attraction.

Efficiency of resources Regional specialisation in crops according to their comparative advantages to improve efficiency of resource use.

Effluent 1. Stream flowing out of a lake, reservoir or larger stream. 2. (industrial) The waste produced from a factory or industrial complex. In many countries, effluent is discharged into nearby rivers or the sea causing extensive pollution.

Electrostatic seed separators These separators can separate seed of different electrical properties, essentially independent of difference in size, shape, weight or surface texture. They separate seed by allowing them to fall free in space and then using an electric field to deflect some seed from their normal fly path.

Eluviation The own washing of clay and other fine particles in free drained soils. The upper horizons are thus depleted of fine clay particles becoming an eluvial zone. The particles may be re-deposited at a lower horizon or may be washed out of the soil altogether.

Elyonuruo hirsu A good fodder grass.

Embryo The rudimentary plant formed in a seed germ.

Emperical Basic principles.

Emulsions as herbicides An emulsion is one liquid dispersed in another liquid, each maintaining its original identity. The two liquids are prevented from reacting with one another by the addition of an emulsifying agent. Ester formulations of 2, 4-D are oil-like and form emulsions. Emulsions are milky in appearance.

Enarching It is a process by which grafting is done by approach.

Endosperm The reserves, stored food of seeds, outside the embryo: referred to as "albumen" by older writers.

Endothal A herbicide. 3,6-endoxahydrophthalic acid.

End point Of jelly To determine as when boiling should cease, is one of the most delicate phases of Jelly making. A test very commonly used is that of dipping a large spoon into the kettle and allowing the juice to run off the edge. When it has reached the jellying point, it will hard from the spoon as a sheet. The motive of sheet will vary somewhat with different juices.

Energy Of a body It is capacity to do work. It is the total work that a body can do. Mechanical energy is of two types (1) Potential, and (2) Kinetic.

Epicarp *See* Plant Science.

Epicotyl That portion of an embryo or seedling above the cotyledons or cotyledony node; plumule.

Eptam Same as EPTC.

EPTC EthylN, N-di-n-propylthiol carbamate. A herbicide.

Equilateral triangle When all sides of a triangle are equal.

Equipment Machines used in doing some scientific work.

Equipment used in seed testirng These are (1) Germinators (*a*) cabinet type and (*b*) room type, (2) Refrigerators, (3) Counting boards, (4) Vacuum counter, (5) Germination boxes, (6) Petri dishes, (7) Germination paper towels, blotters and indented pads, (8) Surgical trays, (9) Wash bottles, (10) Sterilized sand, (11) Table lamps, (12) Sand and water mixtures, (13) Measuring cylinders, (14) Tetrazolium chloride or bromide, (15) Dissecting equipment.

Equipment for seed testing-purity Purity work-boards, forceps, magnifiers, aluminium dishes, seed blower, Chai-o-Matic balance, Torsion balance, other analytical balance, smaller samples pans, and seed and grain sieves. Other things such as racks, working tables etc. are also required.

Essential elements These elements are necessary though in small quantities for proper and healthy growth of plants.

Ethylene In papaya when the latex ceases to be milky and becomes watery, the fruit may be considered ready for harvest. Even when picked while latex is still milky, good quality is said to be secured by using Ethylene to ripen the fruit.

Etiolation Any part of a plant growing in dark has a tendency of losing its pigment.

Etiolation method of grafting In layering when the portion of branch is put under soil, becomes white. This is referred to as etiolation method of layering.

Euphorbia drancunculoides Same as *E. prostrata*, a weed.

Euphorbia hirta Same as *E. prostrata*, a weed.

Euphorbia prostrata Fam. Euphorbiacea. An annual weed. This can be controlled by use of amine 2,4-D at ½–1 lb. a.e./acre.

Euphoribia royleane Also called *Thor* (Hindi). It is an important hedge plant.

European pear *Pyrus communis*. Fam. Rosaceae.

European plum *Prunus domestica*. Fam. Rosaceae.

Evaluation of manures To evaluate a given manure, one ton is taken as equal to 2240 lbs.

Evaporators In countries where the sun is less dependable for drying fruits, specially constructed drying ovens or evaporators are necessary. These make better sanitation possible.

Evaporimeter *See* Plant Science.

Even Uniform.

Evergreen Such plants/trees who do not shed leaves.

Evolution of markets With the more regulated life, markets also started to evolve with greater specialisation such as family market, local market, regional market, natural market, economic market.

Evolvulus alsinoides Same as *Convolvulus arvensis.*

Exalbuminous *See* albuminous.

Exanthema of citrus It is caused by copper deficiency. Multiple buds develop in axils forming short mis-shaped shoots; young twigs die; gum exudes from the base of the twigs. Spray with bordeaux mixture or apply copper sulphate to soil at ½ lb./acre.

Exchange *See* Soil Science.

Exhausted When the nutrients in soils or in plants are all finished.

Exotic Foreign.

Expectation of sample estimates (Sta) It is a property of random sampling that the mean of means derived from random samples approaches the population value as the number of the sample means averaged increased. The same is true of proportions calculated from samples, that is, the mean of sample estimates of proportions is equal to the proportion in the population.

Exports Goods produced by one country to another. In return the country may receive imported goods and services, gold or foreign exchange. The main exports of some of the countries may be machinery, chemicals, motor vehicles, food or textiles.

Extensive agriculture Farming in which the amount of capital and labour applied to a given unit of land is relatively small. *See* Animal Science.

Extrapolation (Sta) The estimating of the value of a quantity or function outside the known range of values. For example, a linear graph showing population growth can be extended beyond the most recent values on the graph. The further from the known range the line is taken, the greater the risk of error in the extrapolation.

Extreme climate A climate in which the seasonal range of temperature is very great. Continental climates are the most extreme as they are found far from the moderating influence of the sea, and in areas such as central Asia the mean temperatures for January and July may differ by up to 50°C.

Eye (botany) The eye-like depression on a underground stem from which vegetative growth starts to form the plant, *e.g.* potato.

F

Factor A gene or character that is being inherited. *See* Plant Science.

Factors affecting growing of fodder crops A good assured rainfall and existence of irrigation facility in dry season favours the cultivation of fodder and forage crops. The scarcity of cultivable land intensely populated areas militates against the production of forage crops in any large area.

Factors affecting machinery maintenance Conditions they are expected to meet in the locality. Owners and operators should have a thorough knowledge of the machines, its operation and maintenance techniques. Thorough study of the instruction books and manuals which are usually carefully prepared and supplied with every machine that had been purchased for a reliable manufacture or his dealer is necessary. Good spare parts' dealership and delivery services are essential.

Fahrenheit scale A temperature scale in which the ice temperature is taken as 32° and the steam temperature is taken as 212° (both at standard pressure). The scale is not used for scientific purposes. To convert between degrees Fahrenheit (F) and degrees Celsius (C), the formula $C/5 = (F-32)/9$ is used. *See also* temperature.

Fall *See* autumn.

Fallow Land that is normally cropped but which is allowed to remain unused over a period of time, in either a tilled or untilled condition in order to restore the fertility. The time period for which it is rested may be as short as one season, *e.g.* winter fallow in high latitudes, or for several years, as in areas of shifting cultivation in the tropics.

Fallowing in dry farming Fallowing a small portion of land by rotation and cropping at once in two years ensures a grain crop even in seasons of low rainfall when the crops of annually cultivated soil fail to set seed. This will ensure food for the family in times of famine. Five or six acres may be fallowed on large holdings.

Family market In this condition, exchanges of goods or commodities are confined only to the members of a family.

Farm diary Diary is very useful for keeping primary and ready record for reference for all happenings, transactions and operations at farm. Besides noting down day to day weather condition, it would be desirable to keep a record of all of farm operations, equipment used, dates of planting and harvesting, yield of crops, amounts paid, income received etc. Besides there are noted of personal interest which a farmer may like to put down in his diary. Entries in register may be made later at leisure from the notes in diary.

Farm expenditure This includes expenditure on (1) manual labour, (2) family labour, (3) hired permanent, and (4) hired temporary labour. Also includes (5) bullock labour, (6) seeds, (7) manure and fertilizer, (8) implements, (9) irrigation charges, (10) land rent, (11) land revenue, and (12) other expenses, if any.

Farm fences Special hay crops like berseem, lucerne or oats are placed in fairly thick layers on farm fences and are raked once or twice before storing. Whatever the method, the cured hay is either stacked as such of baled and stored. Baled hay is easy for transport. *See* Fencing.

Farming hazards The presence of erosion or flood hazards, want of proper drainage or aeration, existence of excessive leaching or a too rapid loss of soil water by evaporation have each an important bearing on the choice of cultivation methods to be adopted or crop rotations to be practised. The provision of adequate organic matter, maintenance of favourable environments for optimum activity of soil micro-organisms and supply of sufficient plant nutrients by use of fertilizers, legume crops and manures are necessary to improve the productive capacity of soil.

Farm income It is the multiple of production and prices, the latter being beyond the control of producer (farmer). It is, therefore, advisable that the farmer gives his more attention to factors which increase production.

Farm manure It is produced by animals. Information as to the amounts produced by farm animals are very useful in fertility calculations. It is generally recorded as tons of manure produced annually by farm animals per 1000 lbs. live weight.

Farm manure as maintainer of soil fertility Farm manure is valuable especially because of the nitrogen it supplies. The organic manure it furnishes may not greatly augment the amount of humus in the soil unless the manurial applications are very large. In spite of the fact that farm manure supplies energy and no doubt has a very considerable influence on the physical condition of the soil and on its micro-organic flora, it must be considered essentially as a nitrogen and to a lesser degree a potash fertilizer.

Farm map A farm map showing the size, shape and location of fields, soil quality, irrigation and drainage channels, approach roads and special physical features, if any, should be shown on the first page of farm account book.

Farm price Such a price is the average price of a commodity at which the commodity is disposed of by the producer to the trader at the village site during the specified harvest period.

Farm production records The production and value of crops grown each year should be recorded.

Sl.No.	Crop	Area	Average Yield	Total Yield	Market Price	Total Value
1						
2						
3						
4						

The area and yields figures may be consolidated from the field records. The crop yields may be valued at market prices prevailing at the time of harvest.

Records of production are useful during the first year chiefly in giving an idea of land utilization and its influence on farm earnings and in affording a comparison with similar information, if available from nearby farms.

Farm records and accounts A farmer may keep some farm records for his own use and as for example for farming is a complex business involving many transactions and day to day decisions.

Farm yard manure Manure originated from the produce of farm come in this category such as dung, compost and green manure etc. Good quality farmyard manure is perhaps the most valuable organic matter applied to a soil. It is the most commonly used organic manure. This is also known as dung manure. The basic material used in making this manure is dung and urine. This should be prepared keeping in view that the manure should be prepared in closed places where it is protected from rain, sun and strong dry winds. A perfectly rotten manure should be used in the fields as unrotten manure may develop many kinds of insects and fungi in the soil.

Fast winds High winds or squalls are experiences during cyclones, depressions, dust-storms, thunder-storms, etc. The mechanical damage to crop plants and trees caused by fast winds is well known. If the air is cold, it causes frosts. On the other hand, if the wind is hot and dry, the rate of evaporation from water reservoirs and transpiration from plants increases enormously. On such occasions, plants wilt or die and the grain in milk-stage shrivels due to rapid desiccation.

Fats and oils Fats and oils are characterized by the readiness with which they undergo hydrolysis or oxidation and become rancid on exposure to air or heat, the process being accelerated under moist conditions. Feeding of rancid oil or fat to animals may cause ententis due to imitation of the mucous membrane of the alimentary canal.

Favourable conditions for micro-flora and fauna Under favourable conditions, such as those obtained after the ploughing in of a green manure crop, the bacteria multiply enormously. On the other hand, in sand desert soils or under water logged conditions, they are but in small number. As between the soil plants, it is estimated that generally bacteria form about 90 per cent of the total population, actinomycetes about 17 per cent and the fungi and algae together about 1 per cent.

Female tree The tree which bears only female flowers (diocious).

Fencing Protection of the trees, or farm of any kind, needs protection from animal pests and prevention of stealing ordinarily requires fencing of some kind, and this is often done to advantage before starting a farm. The fencing should be such as to exclude cattle, goats, deer, wild pigs and donkeys, as well as human thieves.

Fenac A herbicide. 2,3, 6-trichlorophenylacetic acid.

Fenuron 3-phenyl-1. 1-dimethyleurea. A herbicide.

Fermentation The breakdown of organic substance, particularly carbohydrates, under anaerobic conditions. It is a form of aerobic respiration and is seen in certain bacteria and in yeasts. The incompletely oxidised products of alcoholic fermentation ethanol and carbon dioxide are important in the brewing and baking industries. *See* also lactic acid bacteria.

Fermentation of farm manure During his fermentation, the simple nitrogenous compounds are influenced first, although the more complicated constituents are by no means unaffected. Carbon dioxide is given off in large quantities. Urea is readily affected by aerobic activities and quickly undergoes hydrolysis, the ammonium carbonate becomes unstable and produces ammonia etc.

$$CON_2H_4 + 2H_2O = (NH_4)_2 CO_3$$

$$(NH_4)_2CO_3 = 2NH_3 + CO_2 + H_2O$$

Fertility In population studies, the number of live births within given period and area. With mortality and migration, it is one of the three components of population changes in an area. A number of measures are used to compare spatial variations in fertility, the most commonly used measure being crude birth rate, fertility general rate, gross reproduction, net reproduction rate.

Fertility management of mineral soils The maintenance of a profitable and continuous soil fertility is an intricate problem since many variable factors are involved. Weather conditions, moisture relations, soil organic matter and tilth, soil reaction, available nutrients, plant diseases and insect pests are the more important factors that should be considered. *See* farm manure as maintainer of soil fertility.

Fertility and application of lime Three aspects of this problem are first considered (1) shall lime be applied? (2) which form of lime to use and (3) rate of application. These are collectively considered in respect of soil pH, and exchangeable calcium and magnesium present in the soil, etc.

Fertilization and nitrification Nitrification is often known as the weak point of the nitrogen cycle. Soil conditions that affect the vigour of nitrification are (1) soil aeration (2) temperature (3) moisture (4) active lime (5) fertilizer salts and (6) nitrogen-carbon ratio.

Fertilization in flowers When male and female gametes combine to form fruit, grain.

Fertilization of fields *Sea* fertility of mineral soil.

Fertilizer inspection and control Certain provisions are more or less common to the various countries of the world. It is required of the producer to print the necessary data on the fertilizer bag or and authorised tag:

Weight of the package.

Name brand and trademark.

Name and address of manufacture.

Chemical composition guaranteed.

Fertilizer objectives Primary objective in the application of a fertilizer to a soil is to supply plant nutrients, but the addition is usually made to supplement the already present constituents rather than to constitute the only source of supply. The fertilizer should establish a balance between the necessary elements of nutrition. The fertilization of soil flora is an essential feature of successful soil management and fertilization does satisfy their needs first as they have a better adapted to the intake of the nutrients than the higher plants.

Fertilizer Generally the chemicals used to enhance the soil fertility and plant yields. These are given in various compound forms.

It is a substance that is added to soil to make it more fertile by supplying the nutrients required by plants. Natural (organic) fertilizers include manure, bone meal, leaves and decaying vegetation. Modem artificial (inorganic) fertilizers include ammonium nitrate, potassium (from potash deposits) and phosphate. The latter are chemical fertilizers and are mostly used in countries practising intensive cultivation.

Fertilizers and crop management practices Equal attention must be paid to the other soil and crop management practices to ensure good tilth, proper drainage, required soil reactions, soil conservation, good land use, suitable crop rotation, adequate organic matter in the soil and satisfactory soil micro-organism activity. Each one of these plays a vital role in determining eventual production. Neglect of one or more of these factors leads to reduction in yield and creates the need for still heavier manuring. Finally, manuring should not only be balanced in itself, but should also be designed to supplement the good effects of proper land use and beneficial soil management.

Fertilizer mixtures These are formulated products made by mixing together two or more fertilizers.

Fertilizer nutrients and their problems Calcium and magnesium are not rated as fertilizers though important. Nitrogen, phosphorous and potassium are so commonly applied as commercial fertilizers. They are often referred to fertilizer elements. The competition of micro organisms is a difficult one as at times their requirements for potassium become very high. The amount of potassium held in easily exchangeable condition at the colloidal interfaces is usually small. With phosphorous, the problem can be that when added to the soil, it tends to revert to less soluble and less available forms. In fact the practical importance of phosphorous in maintenance of soil fertility is at par with nitrogen and its attendant organic matter.

***Festuca* spp.** These species are susceptible to rust *Puccinia graminis*.

Fibrous root Roots from a plant form in a group on the same plane.

Field 1. An area of land, usually surrounded by hedges, fences or a ditch, which is used for recreation, pasture or cultivation. 2. An area that is rich in a particular natural product such as natural gas, oil, coal, or, gold, *e.g.* oil and coal fields. 3. An outdoor locality in which exploration and research takes place, away from the laboratory, class-room, office, etc. The work done at such locations referred to as field work.

Field capacity The state of the soil once infiltration has seized and all gravitational water has drained down. Water remaining in the soil surrounds particles as film or occupies tiny pores as result of capillarity. Soil moisture that is available to plant roots lies between field capacity and wilting point. Field capacity is expressed as a percentage of the soil's oven-dried weight. *See* capilliarity, wilting points. After a time of full irrigation (one to three days), when the downward movement ceases, the soil is almost uniformly moist. The percentage amount of water held by a soil at this stage, expressed on dry weight basis, represents the field capacity.

Field experiments and fertilizer requirement Field experiments constitute the most reliable method but being time consuming and expensive they are conducted mainly by research farms and research organisations. Farmers wishing to use field experiments as a valid means of determining the fertility status of their soil and nutrient requirement of a particular crop should seek the advice of the state agronomist. Improperly conducted field experiments not only mean waste of time and money but also involve the risk of incorrect inferences and conclusions regarding the most economic fertilizer practices.

Field observations for laboratory studies Field observations are usually supplemented by laboratory studies. Base maps showing accurately the largest number of guide points are employed for the placement of soil boundary lines.

Field output of ploughs and harrows The usual rule of thumb for estimating the output of an agricultural field operation is to multiply the effective working width of the implement in feet by the speed of the implement in miles per hour. The product gives the output in acres for a period of ten hours. This formula allows a reasonable idle time for turns in head land, etc. Thus, a 3 to 12 inch plough being pulled at a speed of 1.5 miles an hour can plough 4.5 acres (3×1.5) in ten hours. It is useful to remember that a speed of 15 m.p.h. is equal to a speed of 22 feet per second. Thus if by a watch you find that a tractor moves 4.4 feet per second it will cover three miles per hour. Average pace of an adult is 30 to 33 inches. It is a good thing to know the length of one's average pace by pacing off a known distance. With some practice one can measure distance by pacing off with sufficient accuracy for all practical purposes. Number of 33" paces to 20 second period divided by ten gives the speed in miles per hour.

Field strip cropping It is the planting of farm crops in more or less parallel strips across fairly uniform slopes but not on exact contours.

Field system The system of farming practised in Britain and parts of Europe during the middle ages. Arable land was divided into strips and members of the community cultivated strips in different parts of the village. Animals were grazed on an area of common land. When the land was cultivated one year and left fallow the next, it is called the two-field system; cultivation for two years and fallow for one is called the three field system.

Field test of seed Since most crops cannot be identified as to variety on the basis of seed characteristics alone, it is often necessary to make tests by growing seeds in the field for varietal identification. Field tests are necessary if it is desired to carefully check the correctness of the labelling of a seed on the seed certification work. Field tests may be used to compare field and laboratory germination results. Samples collected during seed surveys may also be planned in the field for evaluation.

Fig *Ficus carica*, Fam. Moracea. Fig has never been grown extensively in India. Maharashtra formerly had more than about 500 hectares, but largely because of disease, this figure has less than this, and there are a few fig grooves in Hyderabad and in other parts of the country. The types are not of the best quality and it is likely that they are hybrids between the European and other species. The climate of most parts of India seems suitable for fig cultivation. Rain at the time the fruit ripens is undesirable. Various types of soils are used successfully.

Filler Any plant or plants put in the vacant places caused due to failure of any sapling Jo establish after transplanting. In making proper mixtures of fertilizers, sometimes some substances such as sand, ash, soil, ground coal etc. are used as fillers mainly to increase the volume of the fertilizer mixture.

Filling of containers It is necessary that before filling any container with a fruit product, it should be completely sterilised in order that the product is not spoiled in storage and transit before it is used by the consumers.

Filter pump A type of vacuum pump in which a jet of water forced through a nozzle carries air molecules out of the system, filter pump cannot produce pressure below the vapour pressure of water. They are used in the laboratory for vacuum filtration, distillation and similar techniques requiring low grade vacuum.

Financial assistance by agriculture refinance corporation The corporation grants financial assistance only for activities connected with the development of agriculture including animal husbandry, dairy farming, pisciculture and poultry farming. Assistance from the corporation is available for the following schemes:

1. Reclamation and preparation of land so that, in particular, facilities for irrigation or fully utilised.
2. Development of special crops such as areanut, coconut, cashewnut, cardamom, coffee, tea, rubber, etc.
3. Development of mechanised farming, use of electricity through tube-wells, pump sets, etc.
4. Development of animal husbandry, dairy farming, pisciculture (including cooperative fisheries) and poultry farming.

The repayment of the principal and payment of interest on the loan granted by the corporation should be fully and unconditionally guaranteed by the Government.

Finger millet (*Eleusine coracana* Gaertn) Finger millet or ragi is the most important grain crop among millets. It is usually converted into flour, made into cake and porridge or pudding. Malt is also prepared from germinated grain. It is one of the hardiest crops suited for dry cultivation in tracts in which the rainfall ranges between 20 and 40 inches. It can also be cultivated as an irrigated crop.

Fineness of cotton staple The fineness immediately connected with the staple length. The longer the staple the timer will be staple.

Fire fanging of manure A change of biological nature that takes place in a loose and dry manure is called fire-fanging. It looks as if a real fire has taken place in the manure heap while in reality this condition is produced by the fungi and the dry and dusty appearance of the manure is due to the mycelia of fungi, which penetrate in all directions and use up the valuable constituents. Manure thus affected is of little value either as the fertilizer or, soil amendment.

Fish manure It is available as dried fish or as fish meal or powder. In regions where fish oil is extracted, the residue can be used as manure. Depending on the type of fish, its manurial constituents vary from 5 to 8 per cent of organic niteogen and from 4 to 6 per cent of phosphoric acid and 0.7 per cent of potash. It is quick acting and suitable for all crops and soils. It should preferably be powdered before use.

Fixation of nitrogen of atmosphere by nodule. *See* nitrogen cycle, Soil Science.

Flacourtia cataphracta Straw berry; the Indian *paniala* fruit; family Rubiaceae.

Flash point The lowest temperature at which sufficient vapour is given off by a flammable liquid to ignite in the presence of a spark.

Flax A plant with pale blue flowers that is grown for the fibre obtained from its stem and for its seed, which when crushed produces linseed oil. Flax stalks are processed to provide the fibre that is used to make linen, fine writing paper and cigarette paper.

Floods Parts of the world, particularly areas drained by large river systems, suffer sometimes from devastating floods. Rainfalls exceeding five inches in 24 hours occur in practically all such parts of the country and those ranging from 15–20 inches in 24 hours have occurred occasionally.

Flood control The prevention and regulation of flooding. This may be achieved by the building of dams, storage basins and embankments combined with the constant monitoring of water levels.

Flood irrigation In this case, the water from the source is allowed to flow freely allover tile field and is said to be a wasteful type of irrigation as there is no control over the flow or quantity of water.

Flood resistant varieties Another ameliorative measure is the introduction of flood-resistant varieties of crops as done in paddy.

Flow of sap Movement of the vital liquid within the plant.

Flow of irrigation water If the flow of water in the irrigation channels is well levelled, there will be less evaporation losses of water.

Flue curing This is practised in case of tobacco. It has a few stages of drying such as fixing the colour, drying of lamina and drying of the midrib.

Fluid Liquid stage.

Flush Flowering in trees.

Fodder cutters Two sharp steel blades are fixed on a circular fly wheel. An automatic pushing arrangement is also provided. In some areas where power is available, farmers are using the power chaff cutters on rent base. The power driven chaff cutters can be used for cutting dry juar stock also, which is not possible to cut by means of hand-operated chaff cutter.

Fodder yields and good plant stand A good dose of organic manure as basal applicatiqn supplemented by a small top dressing of ammonium sulphate to encourage early plant growth, (2) A relatively heavy seed rate to ensure a dense stand of tall, thin stalked plants, due to lack of tillering, (3) The growth of cereal crops mixed with one or more suitable legumes to improve nutritive value of the fodder, (4) Cultivation of the recommended varieties, (5) Irrigation of the crop to secure optimum plant growth, and (6) Harvesting at the proper stage for feeding green preparation of silage or hay.

Folded bud When the parts of a flower are overlapping each other in the bud stage.

Folded leaf When the leaves in the bud stage are compactly arranged.

Foliar application Spraying of fertilizer or any other chemical or water on the aerial parts of plants is known as foliar application. This is done by the help of sprayer or dusting-machines.

Food chain A series of links by which food energy is passed from organism to organism within a natural community. Green plants (producers) commonly initiate food chains; they are eaten by herbivores, which may in turn be eaten by carnivores.

Food grains Amongst food grains, rice occupies the foremost place, jowar (sorghum), bajra and wheat come next in importance in the order they are named. Maize, barley, ragi and small millets are also grown on a fairly large scale.

Food value of fruits All fruits contain roughly 50 per cent roughage, 8 per cent carbohydrates and 8.5 per cent protein.

Foot rot *See* Plant Pathology.

Forage crops and grasses The agricultural prosperity of a country is intimately bound up with adequate nutrition of the cattle, which provide the principal motive power in Indian farming. Forage crops and grasses constitute an important group in this respect.

Forecast In meteorology, a prediction of the weather for a set period, usually a day, based on observations of present weather patterns and knowledge of the normal development of the systems producing the present weather.

Foreign seeds Seeds of weeds and crops other than the kind under consideration of being tested.

Forest A large area of land covered with trees and undergrowth. It is generally more extensive than woodland; consists of either natural or artificially planted vegetation with tall mature trees, which are often of commercial value; and is dense growing forming a continuous canopy. *See* also coniferous forest, equatorial rainforest, gallery forest in Plant Science.

Forkert budding This method of budding is exclusively used in the Dutch East Indies and has been recommended for such fruits as the mango in India. It is similar to patch budding with the patch about an inch long and a third as wide, but a flap of bark is left at the bottom. In the modified form, the bark is peeled off after making the initial transverse incision, instead of being cut The patch of bark having the bud is inserted under the flap and cover the stripped section of the stock. It is then tried as in other forms of buddings.

Formal garden In which certain scheme is followed while arranging to grow flowers or plants in a garden.

Formalin A mixture of about 40 per cent formaldehyde, 8 per cent methyl alcohol and 52 per cent water. The methyl alcohol is present to prevent polymerisation of the formaldehyde. It is powerful reducing agent and is used as a disinfectant, germicide and fungicide, and also a general preserving solution. In contact with the skin, formalin may cause irritant cermatitis and ingestion can cause abdominal pain.

Formulation of herbicides Formulation refers to the way in which the basic weed-killing chemicals are prepared for practical use. Herbicides are formulated to be applied as solutions of water or oils' emulsions, wettable powders.

Forward market A market in which the purchase and sale of commodity takes place at time 't' but the exchange of the commodity takes place on some specific late date.

Forward trading In order to control the abuses and unhealthy practices in forward trading in agricultural commodities which ultimately reflect on the prices in the primary and secondary markets, a forward contracts (Regulation) Act was passed by the Central Government in 1952. Under this Act, a forward Markets' Commission has been set up whose functions, are both quasi-judicial and executive.

Foundation seed Seed produced from breeder's seed by the original breeder or an agricultural research station. Foundation seed is the source of the certified seed, either directly or through registered seed.

Frame Any structure made to comply to a certain requirement of metal or wood.

Frame working This is an alternative method to top-working of fruit trees. The small branches and side shoots are removed and most of the side shoots replaced by scions of the desired variety. Stub grafting and side grafting are most common methods applied in this operation.

Free energy A measure of the ability of a system to do useful work. *See* Gibbs function, Helmholtz function.

Free fixation of nitrogen The H-ion concentration in itself does not seem to be a factor unless pH becomes rather low. In mineral soils, free fixation of nitrogen begins to slacken noticeably at pH 5.5 and at pH values below 5.0 becomes more or less negligible. Such is not the case with organic soils since they ordinarily contain suitable amounts of active lime even at relatively high H-ion concentration.

Free-flowing seeds Non-chaffy, smooth seeds which will travel easily through sampling or mixing devices.

Free trade area An association of states, the members of which have agreed to trade with one another without import tariffs, quotas, export subsidies or other government measures to regulate trade.

Freezing The process by which a liquid is converted into a solid by cooling; the reverse of melting.

Freezing mixtures Two or more substances mixed together to produce a low temperature. A mixture of sodium chloride and ice in water (–20°C) is a common example.

Freezing point The temperature at which a liquid is in equilibrium with its solid phase at standard pressure and below which the liquid freezes or solidifies. This temperature is always the same for a particular liquid and is numerically equal to the melting point of the solid.

Frequency (Sta) The number occurring in each class is termed the frequency of that class.

Fresh dung It contains 0.3 per cent nitrogen, 1.5 per cent phosphoric acid and 0.2 per cent potash.

Frequency polygon (Sta) *See* graphic representation of frequency distribution.

Frequency curve (Sta) A graph that shows the number of occurrences of values in each of a series of classes. One way of drawing a frequency curve is the first to make a histogram showing the frequency distribution of a set of data. A smooth curve can then be interpolated from this diagram to make a frequency curve. A frequency curve cab be drawn to show the distribution of the annual rainfall of a place over a number of years grouped into millimetre classes on the horizontal axis and the number of occurrences in each class on the vertical axis. *See* graphic representation.

Frequency distribution (Sta) The manner in which the frequencies are distributed over the different classes is called the frequency distribution.

Frequency of hail-storms The frequency of hail-storms is small early in winter but increases generally as the season advances into summer. During the rainy season, the phenomenon is practically absent. The number of days with hail is about ten per year over Himalayas but it decreases rapidly to once in two years over the adjoining plains. The coastal tracts are such regions where hail-storms are comparatively rare.

Frequency of irrigation The frequency of irrigation or the irrigation cycle is determined by the nature of the soil and the water requirements of the crop. The amount of water required to wet the soil to a definite depth varies with its texture.

Frost A weather condition that occurs when the air temperature is at or below 0°C. Moisture, on the surface of the ground and objects freezes to form an icy deposit. *See* also glazed frost, ground frost, hoar frost, rime. This is a serious hazard of fruit crops especially its susceptibility to mango is very serious in the crop fruits.

Frost heaving The disturbance of soil and weathered debris in periglacial areas caused by the freezing of water within them. This results from the water expanding as it freezes. Frost heaving is frequently caused by the formation of large bodies of ice known as ice lenses, which force the overlying material upwards into mounds. *See* also congeliturbation.

Fruit bud differentiation If a tree fails to bear a satisfactory crop, this may be due to lack of flowers or may be caused by a failure of the flowers to develop into fruits. It is thus important that fruit production may be controlled at the time fruit buds are differentiated from vegetative buds as well as when the flowers and young fruits are on the tree.

Fruit butter It is similar to jam but more highly concentrated and has a finer consistency. It is usually highly spiced.

Fruit flakes Mango, banana and guava pulp can be dried into excellent fruited cereal flakes on a roller drier, after mixing the pulp with small amounts of sugar, starch, pectin, etc. In the case of mango, to every 100 lb. of the pulp 10 lb. of *maida,* 5¾g lb. Cane sugar, 5¾ lb. glucose, 1½–3 oz. of mixing of calcium carbonate and sodium bicarbonate (to adjust the acidity of pH to 5.0) are added. The pulp is heated to 80–85°C. The *maida* which is cooked into a plastic cup thrice its weight of water by heating for 2 minutes at 75–80°C. The other ingredients are now added, and the cooked mass homogenized and dried in the form of a crisp sheet on a stainless steel double drum drier using 50–60 lb. steam pressure. The sheet is broken into flakes which have an attractive yellow colour and are rich in 3-carotene (precursor of vitamin A). They can be used in a variety of ways such as in breakfast foods, snack, ice-cream, etc. Tomato flakes can be prepared in a similar manner by taking tomato concentrate of 20° Brix 100 lb., wheat flour 4½ per cent lb., sugar 1 lb., common salt 8 oz., calcium carbonate 1¼ oz., red chilli powder 2 ½ oz. and mustard, cloves, cinnamon, cardamon and black pepper $1^2/_3$ oz. each. These flakes can be used for making tomato soup.

Fruit fly *See* Entomology.

Fruit juices Apples are grated; grapes-stemmed, pine apples, peeled and sliced, pomegranates quartered and seeds removed, oranges peeled and tomatoes boiled gently for about five minutes before extraction of juice. Apple, grape, pineapple and pomegranate juices are extracted by pressing the prepared fruit in a basket press. Orange juice is extracted with a screw type juice extractor. Tomato juice is extracted by screening the boiled tomatoes through a sieve to remove seeds and skin. Cashew apple juice is extracted by steaming the fruit at 5 to 10 lb. pressure for five to ten minutes in a pressure-cooker and pressing the heated fruit in a basket press. The juice thus obtained is strained through cloth and gelatine added at three grains per pound of juice. The juice is stirred thoroughly and allowed to settle for 15 to 20 minutes. The supernatant clear juice is separated by decantation and sugar added to 1 oz. per lb. of juice. The juice may be strained for the second time, if necessary. The juice is heated to about 160°F and filled into previously sterilized crown cork bottles.

The bottles are sealed with crown corks and pasteurized by placing them in water at 160° to 180°F for 25 to 30 minutes and then cooled.

Fruit juice concentrates and powders Fruit juice concentrates, rich in vitamin C, can be prepared from fruit like orange, cashew-apple etc. Orange juice or cashew-apple juice which has been treated with gelatine to remove the stringency, is concentrated to 70–72° Brix under reduced pressure at low temperature. This high density concentrate should be stored in the cold to prevent loss of its high ascorbic acid content.

Fruit paste It is also made much the same as butter, but is dried in sun or with artificial heat until it becomes solid. Less sugar may be used and sometimes dried fruits are added.

Fruit products The preservation of fruits though does not supply a fresh fruit, but through the preserved products can supply fruits throughout the year. The product is more palatable and has many dietary values of fresh fruits.

Fruit rot *See* Plant Pathology.

Fruit scab *See* Plant pathology.

Fruit toffees Real fruit toffees can be prepared from fruits like mango, sapota, guava, jack fruits, etc. by heating the fruit pulp with skim-milk powder, sugar and glucose.

Fruit vegetable This group include tomato, egg plant or brinjal, pepper or chilli and okra, bhindi or lady's finger. The first three belong to the family Solanceae and the last to Malvaceae. They are warm season crops and are very susceptible to frost and cold. The edible portion in these vegetables is the fruit which is usually used when fully ripe in tomato, unripe in egg-plant, both ripe and unripe in chilli, and unripe in lady's finger. These fruits are of high nutritive value and contain vitamins; tomato A, B and C; egg-plant A, B; chilli C and lady's finger A, B and C. Tomato and green chilli fruits are very rich in vitamin C. The pungent types of chilli are used as condiments whereas the large fruited types of chilli are taken raw in salads and cooked. Tomato is taken both cooked and raw and used in preserves, pickles, ketchup and sauce. Brinjal and lady's finger are used or fried as a vegetable.

Fruit vegetable varieties Tomato: Best of all, Marglobe, Sioux, Pusa Ruby, Meeruti, Pusa Early Dwarf, Chhohara (Kulu Valley). A semi-climbing small fruited type rich in Vitamin C and sugar is Pusa Red Plum.

Brinjal: Black Beauty, Pusa Purple Long, Pusa Purple Round, Nurki, Banaras Giant, Sutigota

Manjri Gota, Punjab 8.

French Beans: Chinese Giant, California Wonder, Yolo Wonder, World Beater.

Lady's Finger: Pusa Sawani (Mosaic-resistant), Punjab No. 13, Pusa Makhmali, Padra, T.I. (U.P.).

Functionaries in marketing The transfer of produce or goods takes place through a chain, of middlemen or functionaries. In the primary market, the main functionaries are the producer, villager or itinerary merchant, commission and transport agents. In the secondary market, the processing and manufacturing agents are additional functionaries. Financing agents like shroffs, banks and cooperatives may also take part. In the terminal or export market, the commercial analyst and shipping agent also get involved in the transfer of goods.

Fundamental units The units in length, mass and time that form the basis of most systems, of units. In SI, the fundamental units are the metre, the kilogram, and the second. *See* also base unit.

Ftest (Sta) The analysis of variance table provides a ready means of testing the significance of differences between class means. The comparison of the two causes is done by finding the ratio of the mean square concerned either of the two causes to the errors mean square. The ratio is known as the variance ratio and is denoted by F.

Fungal diseases Any diseases of which the causal organism is a fungus.

Fungicides *See* Plant Pathology.

Furrow irrigation The irrigation water is given through the furrows made in the field.

Fytolon A weed killer chemical.

G

Gall formation The tiny structures on the host plant produced by some insects and also by plant pathogens. *See* Entomology.

Garden It is defined as an adjacent area to a living place, which is decorated with certain valuable, beautiful and good looking plants and thus add to health of the gardener and family.

Garden city Any urban location having a landscape with trees bearing beautiful flowers.

Garlic No distinct varieties exist but types with bold cloves, compact bulbs and thick white covering sheath are preferred.

Gas natural The gaseous form of hydrocarbon containing chiefly of methane that is found in underground reserves of the earth's crust often above deposits of petroleum. Natural gas is formed under the same conditions as oil and originated from decomposition of organic matter. It is used for heating as it gives off great heat when it burns and as a raw material in the chemical industry. Large reserves have been found under the North Sea and these are of great economic importance to the countries around this sea. These reserves are distributed all over the world in the marine as well as ground terrains.

General or mixed market In such markets, a large number of commodities are brought for sale and commodities are bought and sold when buyers and sellers gather together to do the business.

Generative city A city that stimulates the economic growth of the winder region in which it is located (the opposite of a parasitic city).

Germination Resumption of active growth by the embryo in a seed. In seed testing, the emergence and development from the seed embryo of those essential structures which, for the kind of seed in question, are indicative of its ability to produce a normal plant under favourable conditions. It is not mere sprouting of seed.

Germination percentage The percentage of the pure seed of the kind under consideration which produces normal seedlings.

Germination test of seed The ideal test for germination would be to sow the seed in the field, according to normal farming practice but this is not usually practical for, firstly because germination tests are required before sowing time and secondly results should always be reproducible. Germination test must, therefore be made under controlled and standard conditions that cannot be obtained in a field.

Ginning of cotton Separation of seed from the cotton (fuzz) is called ginning.

Ginning percentage In per cent seed the ratio of lint in cotton is the ginning percentage. In cotton in which lint is more in relation to the seed, it is considered to be a superior variety.

Ginger (*Zingiber officinale* Rosx) Ginger is grown for its aromatic rhizomes, which are used both as spice and medicine. It is cultivated in almost all the tropical and sub-tropical parts of India.

Girdler insect *See* Entomology.

Girdling (1) To bring about fruitfulness, it can be induced by increasing the concentration of carbohydrates in the top of the tree without markedly reducing the vegetative activity. In such treatment, the downward flow of the sap is checked. For this, a girdle is made around a branch by either making a cut with a knife around the bark but not in the wood or by making a girdle by means of a wire and encircle it deep in the bark. (2) Sometimes a branch of a fruit tree is incised at the base of a branch by removing a ring ofbark but into the wood. This is done to check the downward flow of the sap.

Glaced fruit It is prepared by dipping the dried candied fruit into a hot, very concentrated sugar solution for about one minute.

Glass house Any structure made of glass around in order to control the outside effects of environment.

GMT Greenwich Mean Time.

GNP Gross National Product.

Gonda *Cordia myxa*. Also known as *Lasora* is a fruit that grown well under arid conditions.

Good ploughing For good ploughing, the following points should always be kept in mind:

1. Furrows should be straight and compact of uniform depth and width, and with even finished ends.
2. Furrow wall and bottom should be clean cut and the top of the inverted furrow slices should be level to give an even seed-bed.
3. Depth of ploughing should suit the soil and the crops.
4. All vegetation, manure etc. should be properly covered.

Good plant species of fodder grasses *Cenchrus* species are specially preferable owing to their high nutritive value; they contain about 10 per cent protein besides large amounts of calcium and phosphorous. The management of this grassland should aim at maintaining a balanced stand of *Dichanthium annulatum* and *Cenchrus* spp. in semi-arid region and *Cenchrus* spp. and *Elyonurus* in arid regions. The deterioration of this cover passes successively through *Cenchrus-Elyonurus* and *Cynodon-Eleusine* stages to the last state represented by *Sporobolus pallidus* mixed with a dominant stand of annual grasses like *Aristida* spp. *Tragus recemosus* and other grasses of low quality species that grow above 10,000 ft. are *Phleum alpinum* associated with *Dactylis glomerata, Poa pratensis, Agropyron longeristatum, Festuca rubra, Danthonia* spp. *Trisetum spicatum* and *Agrostic* spp. pasture legumes like white clover, red clover, midics are quite common in these grasslands. In nutritive value, these grasslands can be considered at par with temperature and alpine pastures in other countries.

Gootee *See* Marcottage.

Gradient The steepness of a sloping surface which may be expressed by the angle in degrees from the horizontal or by measuring the vertical distance and the horizontal distance as a ratio as follows:

Average gradient between two points–horizontal distance/ difference in height.

Grading It means the storing of produce into lots, each of which has substantially the same characteristics with respect to market quality and each bearing its own label or name. These characteristics may be shape, size, flavour, colour, degree of ripeness etc.

Grading of agricultural produce Hitherto grading of an agricultural produce under Agmark was mostly confined to the level of merchants. Very small quantities had been agmarked at the merchants' level ultimately percolate to some extent to the producers as well, it was felt that if the full benefit of grading was to be passed onto the producers, grading of the produce has to be undertaken before sale. This step will greatly contribute towards to payment of appropriate prices to the producers commensurate with the quality of their produce.

Grading of fruits Fruits are graded keeping in view their quality, colour and size. Generally different sizes should be separated so that the packing looks more attractive. Also it facilitates the marketing of fruits making it more convenient for the buyers as what size has to purchase.

Grading and market value In order gain consumers' confidence and establish a rational relationship between the quality of the produce and the price thereof, it is necessary to devote some attention to the proper preparation shifting and sorting of a material before it is taken to the market. This is sought to be achieved by grading the produce in conformity to certain accepted quality standards *viz.*, Size, shape, form, weight and other physical and chemical characteristics.

Graft An artificially induced vegetative fusion or union of parts from different individuals, the rooted part is called the stock, and the part fused to it, the scion.

Graftage The scion and top of the root stock are stems of the same size, they are tied tightly together and waxed. Also called grafting.

Grafting by approach *See* enarching.

Graft hybrid A plant possessing shoots composed of two different genotypes as a result of adventitious buds forming at the junction of the scion and stock in a graft. Such plants are not true hybrids but chimaeras, and any seeds formed possess the characteristics of either the scion or the stock but not both.

Grain A small particle of rock that is smaller than a granule. A grain (*e.g.* a sand grain) is less than 2 mm in diameter. Also the seed of crops is sometime called grain as of cereals, pulsed etc.

Granular herbicides In these formulations, the herbicide is absorbed on, mixed with, or impregnated into a generally inert carrier in such a way that the final product consists of small granular particles. Many carriers are used, including clay, sand, vermiculite and finely ground plant parts (ground corn cobs, tobacco leaf trasnes etc.). Granular materials can be spread by hand or mechanical spreaders. These have advantages over sprays in that water not needed for application; costlier spraying equipment is dispensed with, and the granules fall off the leaves of valuable plants without injury.

Granular structure *See* Soil Science.

Grape (*Vitis vinifera* L.) Fam. Vitaceae (Ampiladeceae). Grape is a sub-tropical crop which grows well in dry climates having short sharp winter and long dry summers. The crop sheds its leaves and rests in winter, puts forth new growth in spring and matures in summer. It does not thrive in regions having humid summers. It tolerates frost during the resting period but succumbs to it readily during its growing period.

Grapefruit (*Citrus paradisi* Macf.) Fam. Rutaceae. The name grapefruit had been derived from its habit of bearing the fruit in clusters like grapes. In India, its introduction is comparatively recent, and its cultivation is confined to a small area. Two other varieties *V. lanata* and *V. ripensia* are grown in the North Western parts of India.

Grape propagation The vine is usually propagated by cuttings.

Grape vine pruning Pruning is usually done in north India once a year in spring before new growth starts. In peninsular India, grape vine is pruned twice a year, once in summer and again about December, the exact period being decided upon the distribution of rainfall. Sometimes girdling or ringing of cane is carried out to hasten maturity and to improve size and quality of berries.

Grape varieties Varieties suitable for different regions of the distinct climatic conditions are:

1. Northern plains: Black Prince, *Bedana*, Foster's Seedling, *Kandhari*, *Dakh* and Muscat of Alexandria.
2. Dry and temperate regions: Thompson Seedless, *Sultania* and *Kishmish White*.
3. Southern India: Bangalore Blue, *Pachadraksha* and *Anabe-Shahi*.

Graphic representation (Sta) The frequency distribution of length of ear of a variety of wheat can also be expressed as graph and indeed this method permits of a ready grasp of certain important features which are common to some types of frequency distribution. *See* frequency curve.

The curve obtained from data, depicts clearly the features of the distribution, namely, that it has the maximum frequencies at the middle of the range and the class frequencies diminish more or less symmetrically in the direction of the two extremes. This curve is called a frequency polygon. *See* frequency curve.

Grassland (natural) A major world vegetation type that is dominated by extensive grassy plains. Natural grasslands occur in both tropical latitudes. The widespread growth of trees in such areas is prevented to a certain extent by low rainfall notably during the dry season. The grasses are adapted to these conditions by temporarily dying back and by long roots that effectively trap available soil moisture. Grasslands merge into forested areas where precipitation is higher and into deserts where it is lower. The temperate grasslands have been largely converted to agriculture, especially for grain production. The grassland ecosystem is structurally simpler than that of forests, having only a field (herb) layer, in which the tallest grasses are dominant. Though soil drought and precipitation levels are important factors in the maintenance of grassland and restriction of forest growth, the grazing by large herbivores and recurrent fire are also contributory factors. Natural grassland is therefore not a truly climatic climax vegetation. *See* also Pampes, Praine, Savanne, Stapppe, Veld.

Grassland management An efficient grassland management aims at maximum utilisation of the forage with minimum interference in the growth and vigour of the chief species. This is generally brought about but adopting improved grazing practices like rotational grazing, deferred and rotational grazing or intermittent grazing with controlled number of animals.

Grassland types Ecological survey carried out recently by the I.C.A.R. has revealed the presence of five distinct grass covers in the country in plains and the hills.

Green gram (*Phaseolus aureus* Roxb.) The cultivation of this pulse is spread all over India. In plant habit, it is more or less similar to. black gram, cowpea and *moth* bean. However, it is more drought resistant than the first two and less drought resistant than *moth*. The green pods are used as vegetable, grains (whole or split) as pulse, and straw and husk as fodder for cattle. Unlike other pulses, it is said to be easily digestible, without producing heaviness or flatulence. Grains are also eaten whole, parched, salted with sugar or boiled with condiments. The crop has a restorative effect on soil.

Green manure The plants or crop selected for this purpose should have deep roots and well branched. The vegetative portions should be succulent and plant quick growing. It should require less of manure for its growth and should be tolerant to scarcity of water. Its seed should be available on cheap rates and the plant should be tolerant to inclement weather conditions. A green manure plant should be resistant to insect and disease attack. It refers to the practice of turning into the soil undecomposed green plant tissue. The ploughing under of some rapid growing crop, such as oats, rye or peas is the example.

The plant should leave a good residual effect on soil. All the parts of the plants should be utilised and the plant should be quick growing as to be ready in short time. It should add to the fertility of soil to a maximum extent and its management should not accrue much of expenses nor it should need much of attention and care while growing. It should have a desirable place in the crop rotation and should be a good seed producer. Roots should have a lot of nodules on them.

Green manuring advantage The practice of green manuring, wherever feasible, is the principal supplementary means of adding organic matter to the soil. This consists of growing a quick maturing crop and ploughing it under to be incorporated in the soil. The green manure crop supplies organic matter as well as additional nitrogen, particularly if it is a legume crop, which has the ability to acquire nitrogen from the air with the help of its root nodule bacteria. A leguminous crop producing three to ten tons green matter per acre will add from 24 to 80 lb. of nitrogen when ploughed under. This would equal an application of three to ten tons of farm manure on the basis of organic matter and its nitrogen contribution. The green manure crops also exercise a protective action against erosion and leaching.

Green revolution The increase in crop production of cereals that has taken place since the late 1960s mainly as a result of the introduction of high yielding varieties of wheat and rice. Through research and plant breeding, scientists produced plants that gave significantly higher yields in some cases doubling the yield and enabling double cropping. The use of fertilizers was also made more efficient.

Greens or leafy vegetables There are nearly 20 leafy vegetables grown in this country. They are grown all the year round. They are easy to grow and are nutritious and rich in vitamins and mineral salts. All these crops are propagated from seed sown direct in the yield where the crop is to stand.

Grey areas Such areas where ground water demand and supply equilibrium is likely to be disturbed.

Gross farm income It is the total value of the farm produce whether sold of disposed of in any other manner, *i.e.* given in kind as wages, consumed at home etc. It includes the money obtained for produce actually sold and the value of the remainder at the harvest rates prevalent in the village.

Gross domestic product (GDP) The total value of goods and services produced by a country over a period of time, normally a year. Goods and services are valued at market prices and goods used for final consumption are included, goods used in the production of other goods, such as steel are excluded. The word gross is used because there is no deduction for replacement capital goods as new machinery.

Gross national product (GNP) This is the gross domestic product of the country plus the income obtained from investment abroad by domestic residents, less income earned in that country from investments made by foreigners abroad.

Ground frost A temperature of 0° or less on the surface of the ground, although the air temperature may be above freezing. It is particularly dangerous to new plants.

Ground layering For inducing rootage in plants, a branch is covered with soil which after sometime starts rooting at this point of soil layer.

Groundnut (*Arachis hypogea* L.) Groundnut is believed to be a native of Brazil (S. America) from where it was introduced into Asia and Africa in the sixteenth century.

The oil-content of the seed varies from 44 to 50 per cent in different varieties. It is one of the chief sources of vegetable oil extensively used for cooking purposes and also for the manufacture of hydrogenated *Vansapati*, soap and toilet requisites. Kernels are also eaten raw, roasted or sweetened. They are rich in proteins and vitamins A, B and some members of the B_2 group. Their calorific value is 349 per 100 gms. The residual oil-cake contains 7 to 8 per cent N, 1.5 per cent P_2O_5 and 1.2 per cent K_2O and is used both as a fertilizer and a cattle feed. The haulms (plant stalks) are fed (green, dried or silaged) to livestock. Groundnut shell is used as fuel for the manufacture of coarse boards, cork substitutes etc.

The crop is also valued as rotation crop. Being a legume with root nodules, it can synthesise atmospheric nitrogen and thereby improve soil fertility.

Ground suckers The saplings that come out from the underground corms of plants.

Ground water The subterranean water is tapped by digging or drilling wells. Wells may be shallow (30 to 40 feet) or deep (86 to 100 feet or more). In either case, it is necessary to lift the water before it can be used as irrigation water. It is generally free of turbidity and bacterial pollution.

Groundnut cake (with cover) It has 4.5 per cent nitrogen, 1.7 per cent phosphorus and 1.5 per cent potash.

Groundnut cake (without cover) It has 7.3 per cent nitrogen, 1.5 per cent phosphorus and 1.3 per cent potash.

Groundnut decorticators When the groundnut is to be sown, labourers are employed for breaking the pods and taking out the grain. In doing so, the upper shell of the grain is broken or the two cotyledons of the grain are also separated, making it useless for sowing because such damaged grain will not germinate. Groundnut decorticators, therefore, have been designed. The Maganwadi type of groundnut decorticator is like a cradle with a screen placed below.

Guanabana Soursop, *Annona muricata*. This can be grown only in tropics.

Growing season The part of year in which plants maintain growth. Temperature is the main factor and the growing season is often defined as the period before lethal frosts. Maize, *e.g.* requires about 150 frost free days for its growing season and about 200 frost free days. As the plants are able to maintain growth all the year round and no growing season can be marked there, the growing season is about two-three months in polar latitudes.

Growth *See* Plant Science.

Guano The accumulated excrement of certain animals, especially seabirds and bats, which can be used as a fertilizer. Large deposits of seabird excrement are found on the Peruvian and Chilean offshore islands where there are desert conditions with little or no rain to wash the material away. Guano is rich in phosphorus and extraction is carefully controlled to conserve the diminishing resources (Spanish). It contains 11–16 per cent nitrogen, 8–16 per cent phosphorus and 2–3 per cent potash.

Guard crops In this kind of mixed cropping, also crops are sown in separate rows. In such mixed crops, one is a main crop for whose safety the other secondary is sown, *e.g.* in a field of sugarcane, sunhemp is sown in 10 to 12 rows all around the field.

Guava *Psidium guajava* Fam. Myrtaceae. It is a very hardy tree, withstands heat and prolonged droughts, but is susceptible to frost. A cool winter induces heavy fruiting. It grows in all types of soils having pH ranging from 4.5 to 8.2. Its fruit is rich in vitamin C (35 to 100 mg. per 100 gm.) content.

Guava cheese Cut guavas into thin slices and boil with an equal quantity of water till soft. Strain the boiled fruit through a sieve or net-cloth. Add 1½ lb. of sugar. 1 gram of citric acid and about 2 oz. of butter for every pound of pulp and cook the mixture to a thick paste. Add a small quantity of a red edible colour and a little common salt to improve the appearance and taste of the produce on a greased plate. When set, cut into pieces and páck in moisture-proof paper.

Guava jelly Wash guavas and cut them into thin slices. Add sufficient water to cover the fruit and add 10 to 15 grains of citric acid per pound of fruit. Boil the fruit for 25 to 30 minutes till the extract becomes thick, and then strain through coarse cloth. A second extract may be taken by adding ¼ of its weight of water to the pomace and boiling. Add 14 to 16 oz. of sugar for every pound of extract and boil till then temperature of the product reaches 221°F. Cool a small quantity of the material in a spoon and allow it to drip. If it drips in the form of flakes, the jelly is ready. Cool the jelly for a few minutes and pour into dry jars. Allow the jelly to stand for 10 to 12 hours to set and seal.

Guava varieties Lucknow 49, Allahabad-*Safeda* and Pusa Seedless are white-flesh varieties. Several types having pink flesh and white flesh with bright red skin are also known.

Guidelines for taking crop mixture Combine the pod and non-pod crops. Mix the trailing crops with tall crops. Mix the shallow and deep rooted crops. Mix shrub crops with tall crops. Mix the crops with the same water requirement. Such crops which harbour the same insects-pests and diseases should not be mixed. Mix such crops which have the same requirement of fertilizers.

Guinea grass *Panicum maximum*. Grown as irrigated commercial fodder under various conditions of soil and climate. The grass is propagated chiefly from root-cutting. Severe cold hinders its growth.

Gully An incised water-worn which is particularly common in semi-arid areas. Lt is formed when a overland flow down a slope especially following heavy rainfall is concentrated into rills which merge and enlarge as gully. Gully erosion of soil and rock takes place by the concentration of runoff into gullies. Also *See* Soil Science.

Gully control Gully control works should be tackled from the ridges and as such the first step should be to plan and execute suitable soils and water conservation works in the catchment area draining into the gully. Proper diversion trenches or perimeter bunds should be constructed around the gully head to prevent ingress of water into the gullies. Suitable measures should also be undertaken to canalise the gully, so as to bring it under cultivation if economically feasible, or grow suitable protective vegetation and plants of economic value. *See* Soil Science.

Gully erosion This is the third stage. When rill erosion is neglected, the tiny grooves develop into winder and deeper channels, which may grow into huge size. This is called 'gully' erosion. Gullies are the most spectacular symptoms of destruction through soil erosion. The gullies tend to deepen and widen with every rainfall. They cut up large fields into small fragments and, in course of time, make them unfit for cultivation.

Gullying Small gullies, while insignificant in the start, soon enlarge into unsightly ditches and ravines. When small enough, such gullies may be ploughed down and seeded with a nurse crop such as oats, barley or wheat, which will check the runoff until the grass obtains a start. Once the crop is established (sod), it may stop the erosion entirely.

Gum arabic A plant product having sticking qualities.

Guntur sour orange Family Rutaceae, *Citrus madraspatana*.

Gypsum Use of gypsum on alkali soils or on black alkali land is sometimes recommended for the purpose of converting part of alkali carbonate into sulphates, thus reducing the injurious effects of the soluble salts. The soil must be kept moist, in order to bring about the reaction:

$$Na_2CO_3 + CaSO_4 = Na_2SO_4 + CO_4 + H_2O$$

H

Hail Occasionally causes damage to many fruits and other crops. Hail is one of the greatest hazards in the apple growing regions in the Himalayas and fruits on the plains also suffer. For most of fruit growers, as well as other farmers, hail is just one of the many factors which make their occupation so much of a gamble. Precipitation in the form of pellets of ice (hailstones) that develop in and fall from a cumulonimbus cloud, either at a cold front or where intense heating of the surface causes rapidly ascending convection currents.

Hail storms These are, however, destructive to standing crops and even to livestock and human beings. Protection against violent hail storms is difficult. But the farmer who is forewarned, can harvest his crops if they are already ripe and keep his livestock under safe cover.

Hammermill A hammermill runs at about half the cylinder speed can remove troublesome appendages. Screen openings should be large enough to prevent damage to the seed, but not large enough to pass the seed out before their appendages are removed. The hammermill should be fed at full capacity to reduce seed damage and to provide the mass needed to rub off the seed appendages.

Hand hoe A small implement to operate on soil manually.

Hand picking *See* Entomology.

Handling and storage of farm manure Since approximately one-half of the nitrogen and three-fifths of potash of average farm manure are in a soluble condition, the possibility of loss by leaching is usually great, even though the manure is not exposed to especially heavy rainfall. Decomposition both of aerobic and anaerobic nature, will cause a rapid waste of nitrogen in the ammonium, nitrogen and elemental forms, one-half of that present in the manure is susceptible to such losses. Packing and moistening the manure will change the decay from aerobic to anaerobic, thus reducing the waste of gaseous nitrogen while encouraging the simplification of the manurial constituents. Tight floors in the sheds/stables/pens and impervious bottoms in the manure pits or under the manure pile should considerably reduce the leaching losses.

Hand shear also called scateur It is a cutting tool of scissors shape and is used as a pruning tool.

Hardiness Capacity of organisms to tolerate intense weather conditions.

Hard seed Seeds which remain hard at the end of the prescribed test because they have not absorbed water, owing to an impermeable seed coat. This occurs in legumes and a few other ends.

Hard wood The portion under the bark.

Hariq (Arabic) The burning of wild grasses to fertilise the ground just before a crop is sown. It is practised in Sudan where the wild grass has been allowed to grow as part of the rotation pattern.

Harm caused by weeds Weeds reduce crop-yields on account of their competition with crops for water, soil-nutrients and light. Certain weeds may further reduce crop growth and subsequent yield by releasing inhibitors or poisonous substances into the soil. They increase the cost of labour and equipment, render harvesting difficult, reduce the quality and marketability of agricultural commodities, harbour insect, fungal, virus and bacterial organisms and some are poisonous to human beings and livestock.

Harmful alkali salts Sulphates and chlorides of sodium, calcium, magnesium and potassium, when concentrated in the A horizon of an aridic soil or on the surface as a white incrustation, are common in alkali regions especially during dry periods. A salt mixture of this character is called white alkali.

Harrow plough The harrow ploughs are lighter and have a large number of small diameter discs for the same width of cut.

Harrows These are equipped with discs or spring types or spikes. Disc harrows are used mainly for tillage operations on heavy or hard soils, but they can also be used for a variety of other jobs such as covering of seed and destruction of weeds.

Harrows and cultivators The indigenous harrows and cultivators are known as *bakhars* and *guntkas*. They mostly consist of a long sharp blade fixed to a wooden implant. The body of these implements is made up of babul wood, cut into a rectangular or octagonal shape. The length of the body is 3′ to 6′ and the width is 10″ to 12″. The blade is attached to the long on its underside by means of two slanting wooden pegs. These implements are used for preparing seed bed, for covering the seed as well as for the summer cultivation, provided the soil is not too hard.

Harvest price Same as Farm price.

Harvesting Plucking of ripe and mature fruits or cutting the crop clean of the field is a matter of great personal experience. In any case fruits should be harvested keeping in view that during the time of plucking them and its availability to the consumers, fruit should not be spoilt and its taste, flavour should be fully well developed.

Harvesting implements So far harvesting operation has not yielded to any mechanised tool. Almost universally, sickle is the only tool that is used for harvesting, even though efforts have been made for introducing reapers, mowers and even combined harvesters and threshers. These have, however, proved successful on big estates or Government farms, but because of their cost they have not become popular with ordinary Indian farmers. The European scythe has also been tried but except for cutting of lucerne or some such green fodder crop, they have not proved very useful. One of the main reasons seems to be shattering of the grain due to the impact of the implements. In the case of sickles, there is less shattering. The sickles are of two types, namely plain or serrated or saw-shaped. The serrated sickles are used for cutting of paddy.

Harvest of crops To gather the produce and end produce from the plants in the fields.

Harvesting rain water Rain water is collected by artificial means as by digging lakes so as to increase the ground water level of the land.

Head of flower Same as Capitulum. *See* Plant Science.

Heading back To check the forward of tall growth in a tree. This is done by cutting off the advancing shoots or twigs in a tree so that instead of vertical growth of the tree the horizontal growth is encountered.

Head system of grape vine training The vine is trained in the form of a dwarf bush. It is allowed to grow into an upright stem with the help of a support and the developing shoot cut off at a height of 1 to 1.25 m in July. It is allowed to develop three to four lateral branches, each about 0.45 m long, arranged in the form of a vase or globlet. At the first dormant pruning in February, the lateral branches are shortened to spurs with one to two buds each. These spurs provide suitable arms for the framework during the succeeding summer. At the second dormant pruning, eight to ten arms, with one to two spurs on each arm, are retained

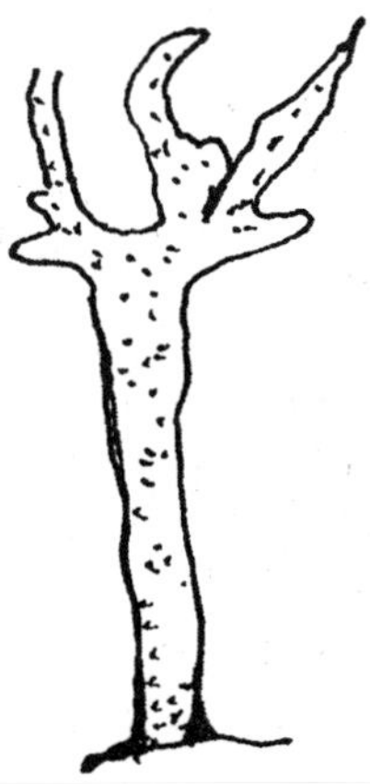

A Grape Vine Headed Back

for cropping in the third year. This system is cheap, but yields are poor.

Head system of irrigation In this system, the subsidiary channels are made to start from a main channel.

Heating of garden Heating the garden by lighting small fires helps to raise the temperature by as much as 10°F in some cases. Smoke screens are not effective as smoke does not absorb all the outgoing heat radiation and re-radiale it to the ground. Protection of orchards from frost on still nights is often practicable, although the most effective methods may seldom be economical.

Hectare symbol ha A metric unit of land area, equal to 10000 square meters.

Hectare symbol h A prefix denoting 10^2. For example, 1 hectometer (hm) = 10^2 metres (m).

Hedge A group of plants used to either demarcate the boundary of orchard or for preventing the animals from entering the field or for ornamental purposes or for use as a fence to check the undesirable animals from entering the hill, or thickness. They are also called the living walls or boundary of an orchard or field. Some of these are useful in flower gardens where the purpose is to form an artistic border or division. Strong thorny hedges are more desirable for orchards. These are species of Acacia and Citrus, and the karanda are the best.

Heliotropium eichwaldi Fam. Boraginacea. It is an annual weed. Can be controlled by Simazine 2–4 lbs./acre in maize and sugar cane, is lethal to weeds.

Hemp One of a number of plants that produce strong coarse fibres. True hemp (*Cannabis sativa*) is an annual plant, which grows to a height of 2–5 m. The fibre is obtained from the stems and used for making rope, twine and matting. The plant also produces the narcotic drugs hashish and marijuana. Other plants producing hemp are known as tropical (or hard) hemps and the fibre is obtained from their stems.

Herbicides These are chemicals which are used in control of weeds. They are of selective composition and therefore should be used with caution.

Hexagonal system of planting In the equilateral triangle or the hexagonal system, trees are planted at the corners of the equilateral triangle. The main advantage of this system is that using it about 15 per cent more trees can be planted per acre at any given distance.

Hibernation *See* Entomology.

High farm-out-turn It is difficult to increase income from the farm. But there are ways to increase the income and the first consideration should be that high out-turn means increased volume of sales and lower unit costs. All efforts should be canalised towards the increase per acre yield (much above average yields) by adequate use of water, manures and fertilizers and by protecting crops from the depredations of pests and diseases.

Hill placement *See* Band placement.

Hinge budding In this method of budding, the cut is made in H-shape in the bark of the stock. The bud is taken on a rectangular shaped patch and inserted into the bark of the stock and tied.

Hinterland The area that has close economic, social and cultural ties with a central place, such as a port, city or town. In the case of a part, it is the region inland from which it receives trade or to which it sends goods. Cities and other settlements have trade areas surrounding them.

Histogram A type of graph that represents frequency distributions by means of rectangles or bars. The widths of the bars are the class intervals, the heights are the frequency. Histograms are frequently used, for example, to show the incidence and amounts of rainfall over periods of time. *See* also frequency curve.

Hitching the plough This is done to adjust the depth of ploughing by adapting a suitable slope of the beam with the yolk.

Hoeing The field operation done by help of hoe in order to break the upper crust of the soil and simultaneously clean the soil of existing weeds and stones etc.

Hoeing in rain-fed crops In the rain-fed crops, hoeing by bullocks is done at the most three to four times, after which the plants grow high and bushy and also the soil becomes hard. To cope up with this condition stronger and sharp hoes have been designed. These are commonly known as the Washim hoe also called the Akola hoe, the Baroda hoe and the Trifali. The design of these hoes are more or less similar, consisting of three tines with a sharp-triangulate blade fixed to each. These hoes are very useful for incultivation in rows of cotton crop.

Hoeing of line-sown crops For hoeing line-sown crops by means of bullock power a light desi plough is used. In the area between the Narmada river and the Krishna river, multi-tined bullock operated hoes are used. These are known as *doura, douri, dundia* and *dulari.* These are miniature forms of *bakhars*. When the plants are small multi-tined hoes fixed to a single wooden body are also used. The draft of these hoes is very light and sometimes two to four of them are pulled by one pair of bullocks. In the dry farming areas, a *doura* with a split blade is used. The blades in this case work in either side of row of plants.

Hole *See* planting hole.

Home garden *See* kitchen garden.

Home mixing of fertilizers If a ready-mixed fertilizer is not available in the market or is too expensive, the separate fertilizers may be purchased separately and the mixing done on the farm. This is home-mixing. *See* mixing of manures.

Homogeneity test A test to measure variation within a lot of seed.

Hopper insects *See* Entomology. Such insects which move about by hopping.

Horticulture The cultivation of flowers, fruit or vegetables using intensive methods of farming. The most intensive form of horticulture is probably the cultivation of crops such as tomatoes or cucumbers under glass in a controlled environment. Less intensive forms include the cultivation of daffodils in small fields in Cornwall and the Sicily Isles and fruit growing around Evesham and in the Vale of Kent.

Horticulturist A person who cultivates, produces and markets horticulture products like fruits, vegetables and their by-products.

Horse gram (*Dolichos biflorus* L.) Horse gram is extensively grown pulse. It is grown mainly to furnish feed and fodder for cattle and horse. It makes excellent hay and is suitable as green manure.

Host plant A plant which allows the insect pest or diseases to live on it for their survival.

Hot pack method of canning The fruits after cut in convenient sizes are boiled. After they are filled in the can and hot syrup is poured in the can.

Humidity The amount of water vapour present in the atmosphere. This may be expressed in several different ways. *See* absolute humidity, relative humidity and saturation.

Hurricane A wind that reaches force 12 on the Beaufort wind scale, *i.e.* it has a velocity in excess of 32.7 m per second. It is destructive to agricultural and horticultural crops.

Hybrid The first generation offspring of a cross between two individuals differing in one or more genes; the progeny of a cross between the species of the same genus or of different genera. The offspring of parent differing in species or the genetic constitution. In the initial stages, the heterozygote was also called a hybrid.

Hybridization The crossing of individuals of unlike genetic constitution.

Hybrid vigour *See* heterosis. The increased vigour, growth, size, yield or function of a hybrid progeny, over the parents, the increase in vigour or growth of a hybrid progeny in relation to the average of the parents.

Hypocotyl The part of the embryo or seedling below the cotyledonary node and above the root the transition region connecting the stem and root.

Hypogeal Plants in which the cotyledons remain below the surface of the soil at the time of germination.

Hypsometer A scientific instrument that is used to calculate altitude or atmospheric pressure by measuring the temperature at which water boils. As the boiling point varies with air pressure, this can be calculated from the temperature at which boiling occurs. An equation is then used to calculate the altitude. The accuracy of this method is not very high.

I

Ilame A variety of genus *Annona* named *A. diversifolia;* is an edible fruit.

Immune A variety of a crop that hundred per cent resistant to certain diseases and insect pests is termed immune.

Immunity The ability of plants and animals to withstand harmful infective agents and toxins.

Immunostimulant A chemical, drug, stresser or action that elevate the non-specific defence mechanism or the specific immune response. These are given by themselves to activate non-specific defence mechanism as well as heightening a specific immune response. It offers a wide range of attraction methods for inducing and building up protection against various such pathogens, bacterial parasites, fungal and viral. Substances like glucose, ascorbic acid and livamisole have tremendous potential in a sustainable fish farming.

Improvement of agricultural production A few items have been taken up under this head, such as–(1) Cultivation of fallow and waste lands, (2) Use of improved seeds and manures, (3) Soil conservation in order to prevent soil from deterioration, and (4) Improving irrigation. All these factors add to the agricultural production.

Inarching Same as attached method of grafting. Also *see* vinier grafting and attached method of grafting.

Inbred line A line showing a considerable degree of homozygosity as a result of continued inbreeding and selection.

Incomplete fertilizers Manufactures provide only two or sometime only one fertilizer element in the brand. Such brands are the incomplete brands.

Incomplete manure Such manures which can satisfy the need of one or two nutritive elements to plants are incomplete manures and include some of the chemical fertilizers.

Increased production due per unit

Sl.No.	*Manure*	*Value per Unit*	*Increased Production Value Due per Unit*	*Net Profit*
1.	Ammonium sulphate	Rs. 4/-	Rs. 10/-	Rs. 6/-
2.	Calcium ammonium nitrate	Rs. 5/-	Rs. 14/-	Rs. 9/-
3.	Ammonium chloride	Rs. 6/-	Rs. 12/-	Rs. 6/-
4.	Sodium nitrate	Rs. 7/-	Rs. 11/-	Rs. 4/-
5.	Sodium chloride	Rs. 8/-	Rs. 13/-	Rs. 5/-
6.	Urea	Rs. 5/-	Rs. 13/-	Rs. 8/-

Increase of soil nitrogen by legumes If the soil is already high in nitrogen or if the crop is harvested and removed, soil has lost some of its nitrogen. Beans and peas are specially likely to leave the soil depleted since fixation is low and their roots often are partially removed during harvesting.

Increase of soil nitrogen due to green manure If a legume crop is turned under as a green manure, there is a possibility of increasing the soil nitrogen to the extent of symbiotic fixation. This is important because the organic nitrogen of the soil increases resulting in increased soil fertility as well as addition of humus in the soil.

Indented cylinder separator This machine used in seed processing is a rotating, almost horizontal cylinder with a movable, horizontal separating trough mounted inside it. Thousands of half-round recesses or indents line the inside surface of the cylinder.

Indian corn *See* maize.

Indian fruit industry Area under fruits was 3.94 million hectares in 1993–94 as per FAO statistics, which is second after China which grows 33.23 million tonnes of fruits.

Indian kaki ***Diosperos embryopteris*** Fam. Ebnaceae. It is said to contain about 50 per cent pectin which prove a good sizing material for textiles.

Indicator species An organism that can be used to measure the environmental conditions that exist in a locality. Lichen species are indicators of level of pollution, as different species are sensitive to different levels and types of pollutants. Tubifex worms indicate low levels of oxygen and stagnant water.

Indigenous Denoting an organism that belongs in or is native to a certain locality or habitat *i.e.* it is not introduced. This description may be applied to species or communities of plants and animals and to human populations. It is sometimes applied to rocks and minerals that have remained in the location in which they were formed.

Indore ridger This implement has a sharp beam with sloping ends and hexagonal share attached to it by strong iron bar. The ridger can be used up to a later stage of crop development without doing physical injury to the plants. It not only eradicates weeds but also produces a good mulch. Similar to the use of doura, two or three such ridgers can be attached to a pair of bullocks.

Indole acetic acid (IAA) A naturally occurring auxin. *See* auxin.

Inducible enzyme *See* adaptive enzyme.

Sl.No.	*Designation*	*Chemical name*
1.	Dalapon (Doupon)	2-2 dichloropropionic acid (sodium salt)
2.	Diuron	3-(3, 4-dichlorophenyl) 1, 1-dimethylurea
3.	Fenac	2,3,6-trichlorophenylacetic acid
4.	MCPA	2-methyl-4-chlorophenoxyacdetic acid
5.	MH (maleic hydrazide)	1, 2-dihydrophridazine-3,6-dione
6.	Simazine	2-chloro-4, 6-bis (ethylamino)-S-triazine
7.	2,4-D	2,4-dichlorophenoxyacetic acid

Informal garden It is also a natural system of gardening. It adds to more enjoyment to the viewer of the garden. It does not follow any scheme of arrangement of beds and flowers.

Infrastructure of market Facilities and services prevalent in a market.

Infrastructure support Such requirement as to provide cold storage, transport facilities, also credit support market information and insurance to fully exploit opportunities.

Inga duleis A strong hedge plant.

Inorganic ammoniates These constitute the inorganic sources of ammonia and nitrogen:

Sodium nitrate $NaNO_3$	Chile saltpetre and synthetic	16% Nitrogen
Ammonium sulphate $(NH_4)_2SO_4$	By-product from coke, gas and synthetic	22% N
Calcium cyanamid $CaCN_2$	Synthetic	35% N
Ammonium nitrate NH_4NO_3	Synthetic	26% N
Ammonium chloride NH_4Cl	Synthetic	15% N
Calcium nitrate $Ca(NO_3)_2$	Synthetic	42% N
Urea CON_2NH_4	Synthetic	16 to 20% N
Cal-nitro $CaCo_3$ + NH_4NO_3 Ammoniated superphosphate	Synthetic	3 or 4% N
NH_4 H_2PO_4 and $(NH_4)_2SO_4$ plus calcium phosphates	Synthetic	(16–18% P_2O_5) 30% N
Ammo-phos NH_4, H_2PO_4 and other ammonium salts	Synthetic	(48% P_2O_5) 11% N
Potassium nitrate KNO_3	Synthetic	(44% K_2O) 15% N

Inorganic manures These manures can give only one or two nutrient elements to the plants. They are very rich in the elements present in them. These manures do not increase the water absorption capacity of soils. Nutrients are readily available through them and their effect in the soil does not stay for long. Their use for a long time endangers the physical condition of the soil *e.g.* prolonged use of ammonium sulphate makes soil acidic and use of sodium nitrate for a long duration makes a soil alkaline. Under both these situations, the chemical and the physical condition of the soil deteriorates. Their use does not guarantee any profits rather sometimes a real loss can be incurred. They have no effect on soil aeration and their prolonged use does not increase the microbe population of the soil. They do not contain oximones. The nutrient elements present in these manures become readily available to the plants. Their use also prevents a balanced growth in the plants *e.g.* more nitrogenous manure increases the vegetative growth in plants and in contrast phosphatic manure lessens this fault to a great extent. Their use also makes plants strongest and such plants produce more flowers and fruits. They do not have any influence on soil temperature. Their use increases the water requirement of crops. Their prolonged use makes soils infertile and to the extent that the soil is rendered completely useless for growing any crop. Carbon-nitrogen ratio of the soil is not affected by their use. Thus all manures having no organic matter in them come in this class and are all chemical fertilizers.

Inorganic substance absorption As per electrolytic dissociation, many compounds when in solution break up into electrically charged portions called ions. Ions may be a single atom or groups of atoms.

Input Whatever is added to the crop during its production *e.g.* fertilizers, irrigation, insecticides and fungicides etc.

Input-output analysis The measurement of the relationships between the level of production of a particular set of goods and services (output) and the raw materials, labour and other manufacturing inputs required to produce at the level of output.

Instability, atmospheric The condition that occurs in the atmosphere when an air mass has a higher temperature than that of the air surrounding it. Such an air mass tends to rise, expand and cool. This is one of the commonest causes of rainfall, which is often heavy and accompanied by thunder.

Instrumentation The measurement of the condition and the control of processes within a chemical plant.

Insular climate The climatic conditions experienced by island and coastal areas where the sea rather than a land mass has the major influence. The main effect of this marine influence is a small seasonal temperature range (*i.e.* equable conditions). This is a direct contrast to the conditions found in a continental climate.

Intake The point at which air of a liquid is channeled into a plant or a mine. For example, surface intakes allow fresh air to be drawn through an underground working.

Intensity of rotation It is found by the formula:

$$\text{Intensity of rotation} = \frac{\text{Number of crops in rotation}}{\text{Years of the rotation}}$$

Intensity of rainfall A measure of how heavily rain has fallen. This can be described generally *e.g.* light drizzles, torrential downpour. More precisely, the hourly intensity of rainfall can be measured by dividing the total rainfall by the number of hours of rain. A daily intensity covering a longer period of time can be obtained by dividing the total rainfall over a set period by the number of rain-days.

Intensive cultivation Methods of farming in which large amounts of capital and/or labour are applied per unit of land. In advanced societies, machinery, modern irrigation techniques, fertilizers and other capital intensive methods are used. In some Third World countries such as Indonesia, intensive agriculture involves the use of a large labour force rather than capital to produce rice and vegetables.

Intercrops To make better economical use of land, it is generally useful to grow some quick growing vegetable, cereal or pulse crops in intervening period of planting and bearing stage of fruit trees. But this one can have some earnings during the period. It should be advised that whatever intercrop one has to grown, its cultural requirements should be different from the main crop. All the intercops are only secondary crops.

Intercultural implements Most of the interculture is done by means of hand tool called *khurpi*, which consists of sharp-edge triangular or sickle shaped blade. Sometimes weeding is done by small spade hoe or mamoty. Most of these tools have small handles and users have to squat while doing the weeding.

Weeding is said to be one single operation which can increase crop yields without much additional investment because efficient weeding will release moisture, crop nutrient and to some extent air and light required for the crop. To increase the efficiency of weeding, long handled hoes such as the Singh Hand Hoe, the Sharma Hand Hoe and the Dutch Hoe have been introduced.

Inter-row spacing Space between any two rows.

Integrated seed testing programme A seed law without some provisions for unbiased seed testing basis loses its value. On the other hand, seed testing laboratories, which are not properly integrated into a seed certification and seed control program, have questionable value also. Both aspects are needed and must be linked together to achieve a properly integrated effective quality oriented seed testing programme.

Integrated pest management Deep ploughing, early sowing of crops, removal of off-type plants (weeds) from the fields and bunds, collection and destruction of picked living larvae, make a trench around the field to check incoming of those insects pests. Dust the crop with Phosobon or Methyl parathion 2 per cent, Malathion 5 per cent or Endosulphan 2 per cent dust @ 25 Kg/Ha or spray crop with Endosulphan 0.07 per cent or Monochrotophos 0.02 per cent, use of neem seed extract, neem seed kernel extract or neem oil (Margosa oil) 16 mi/litre also helps to control many insect pests of the various crops.

Internal drainage (inland drainage, interior drainage) A drainage system in which waters of rivers empty into an inland sea or lake without reaching the ocean.

Internal rate return (IRR) The discount rate in which the present value of a future income from a project equals the present value of total expenditure (Capital and Annual cost).

International certificate The orange colour certificate covers a test of an official seed sample, drawn and sealed by an authorised person. The blue certificate is used where the orange cannot be applied with such as on a submitted service sample.

International seed testing rules' general procedure It prescribes the various techniques to be used in sampling to provide a representative sample from a lot of seed to be certified.

International seed testing association This body has developed some standard procedures which are expected to be guidelines for all the-laboratories in the world. However, as per need of the climatic zones especially of tropical seeds, these procedures are to be worked out. This is also based on scientific experimentation and results.

Interval of irrigation The interval at which a crop requires to be irrigated can be calculated by knowing the daily water requirement of a crop and the type of soil. For instance, the daily water requirement of cotton is 0.21 acre-inch. On a loamy soil which requires 1.33 inches of water to wet it to one foot depth, the interval is six days (1.33–0.21). Therefore, for cotton growing in a loamy soil, it is necessary to apply 1.33 inches of irrigation water every six days to wet it to one foot depth, presuming there is no rainfall. Depending upon the amount of rainfall received during an interval, an adjustment in actual irrigation water to be applied can be made.

Introduction In biogeography, a plant or animal that has been introduced by man into a locality to which it is not native. Important food plants, such as the potato in Europe and commercial crops, such as rubber trees in Malaysia, have been past introductions. Some pests, such as the rabbit in New Zealand, have spread after deliberate or accidental introduction.

Introgressive hybridization (introgression) The introduction of genetic material from one gene pool to another by hybridization and subsequent backcrossing to one or other of the parents. It forms the basis of most livestock and crop-breeding programmes.

Inventory It is a classified list of all assets and liabilities of the farm business at a given time, together with the value of each item. It shows what a farmer owns and what he owes to others.

Inventory format It is the basis for keeping farm accounts and is very useful record of property. To be of maximum value, it should be itemized, showing the number and value of each item under land, buildings, means of irrigation, livestocks, equipment and machinery, farm produces, feeds and supplies.

Invisible exports Those items, such as services, provided for foreign countries that produce income but do not involve the transfer of goods.

Ipomea repentens Family Convolvulaceae. Same as *Convolvulus arvensis*. It can be biennial or even perennial also. A weed plant.

IPC A herbicide. Isopropyl N-phenylcarbamata.

Iron bacteria Filamentous bacteria of the order Chlamydobacteriales. They grow by oxidizing ferrous compounds to ferric oxide. An example is Leptothrix.

Irrigation The artificial application of water to the land in order to grow crops or to improve crop yields. Where large amounts of capital are available, concrete dams and channels are built. A flexible system, which is increasing in popularity, is the use of mobile overhead pipes, which can be moved from field to field. This operation of supply of water to plants depends on the atmospheric humidity, rainfall, season and type of fruit trees. The water holding capacity of the soil and texture and structure of soil decide the intensity of irrigation to be given to trees and crops. The quantity of irrigation water also depends on the type of fruit trees. Experience of the orchardist or crop grower decide the amount of irrigation water to be given to the crop or fruit trees. The various systems followed are–(1) Check or bed method, (2) Basin system, (3) Ring system, (4) Furrow system, and (5) Flood system.

Importance of irrigation For successful farming, irrigation is necessary in one form or the other in all parts of the country where the mean annual rainfall is less than 30 inches. It is indispensable for the economical use of land. When irrigation is provided, both land and labour can be put to profitable use throughout the year, crop yields can almost be doubled and money-return increase by growing valuable cash crops.

Irrigation and depth of soil As the root zone for most crops is usually deeper than one foot, the exact frequency of irrigation can be arrived at by determining the depth of soil from which moisture is being extracted by the growing crop and the moisture content throughout is at the depth.

Irrigation during cold waves Irrigating the field is only useful during cold waves of short duration, as during protracted spells, a considerable part of sun's heat is lost by simultaneous evaporation from the wet soil which counteracts the protection afforded by irrigation. As the air temperature a few feet above ground level is appreciably higher than near the ground, forcibly mixing the two with the aid of powerful fans also helps to raise the temperature of the cold air at ground level.

Irrigation in relation to crop The available water held by a soil is not equally availed of by the growing plant. At the field capacity stage, the plant draws more water from the soil and grows rapidly. The growth, however, progressively decreases as the soil moisture approaches the wilting point. In practice, the crop should never be allowed to extract more than 60 to 70 per cent of the available water. If the water is allowed to be reduced below this level, the plants wilt temporarily and the growth is checked. Though the maintenance of available water at a high level (field capacity) ensures maximum vegetable growth, the quality and sometimes the quantity of marketable produce suffer especially in the case of fruit and seed crops. In the case of vegetables, fodders and other garden crops, a high moisture level helps to develop succulence and tenderness. Young plants should, however, be not allowed to suffer for want of water and should be irrigated more frequently as they have very small root systems. Plant roots grow deeper in soils that are kept moist, but not wet, during the early stage of growth. As the roots grow deeper, a greater depth of soil requires to be kept moist.

Irrigation in relation to soil When irrigation water is applied to dry land, it wets the soil which soon gets saturated as more water is applied. At this stage a part of the water is absorbed by soil particles and the rest is present in the soil pores from which soil air has been driven out. The free water in the soil pores moves down into the lower dry layers and wets the soil in those layers.

Irrigation-soulets *See* water courses.

I.S.T.A. International Seed Testing Association.

Italian millet (*Setaria italica* Beauv) Italian millet or Foxtail millet is cultivated in many parts of the world. It matures quickly in 90 to 120 days giving almost as good yield as crops of much longer duration. It can be grown even at an altitude of 6,000 feet. Alluvial loams, red loams, light soils, black cotton soils and clayey soils are all suited to it. The crop is raised both under rainfed and irrigated conditions and can be grown in hot climate throughout the year.

J

Jack fruit *Artocarpous integrlifolia* and *A. heterophylla*. Jack fruit is grown in South India as stray trees in home gardens and coffee plantations where it flourishes in the humid climate on hill slopes. Warmer plains are suitable provided there is adequate soil moisture. Cold and frost are harmful.

Jacobsen germinator A special device for seed testing in which water is supplied to blotting paper substrate by means of wicks.

Jafna method of curing of tobacco This method is called fire curing. This is mainly used for chewing tobacco.

Jams Prepare the fruits as for canning and cut them into small pieces. Boil the fruit with a little water to make it soft. For every 4½ lb. of prepared fruit, add 5½ lb. of sugar and about ¼oz. of citric acid and cook to a temperature of 221°F or until the product thickens to the consistency of a jam. The addition of sugar may vary a little in accordance with the maturity of the fruit.

Jamaica Also known the bluegrenads variety of Banana. It was introduced in Jamaica from Martinique by a French Botanist at planter, Pouyat in 1836 and soon became very popular.

Jamun Also called Jambolan. *Euginia jambolana*. Fam. Myrtaceae.

Japanese gardens These are natural gardens constructed with a matching layout of the environment and landscape.

Japanese plum *Prunus salicinia*. Fam. Rosaceae.

Jujube (*Zizyphus jujuba* Mill) Fam. Rhamneae. Jujube, ber or bor, is estimated to occupy about 4,000 hectares mainly in the northern, eastern and western parts of India. Almost any soil is suitable, even if it contains enough alkali to prevent the growth of some crops. The tree can stand prolonged drought and also water-logging. It cannot stand frost.

Jujube propagation Propagation is mainly by seed. Shield and ring budding in situ are commonly employed in the North. Jharberi or seedlings of the same species make a satisfactory rootstock. They may be planted six meters apart. Seedlings of jujube trees may be top worked by cutting the trees back half to one and a half metres from ground and budding the new shoots in spring or autumn. Though very little care is given, proper manuring, clean cultivation and irrigation while the fruit is developing help to give increased yields.

Jute (*Corchorus capsularis* L. and *C. olitorious* L.) As a best fibre, it is obtained from crops as raw fibre. The plant grows to a height of about 3m and the inner bark produces fibres of about the same length. Most of the world's jute comes from the lower Ganges-Brahmaputra valley in Bangladesh and India.

The fibre is used for manufacturing packing cloth, hessian and bags for the transport and storage of cotton, wool, foodgrains, pulses, etc. Jute is also used for making rugs, carpets, curtains, upholstery, linings, ropes and twines. Green leaves of *C. olitorious* are edible. Stalks stripped of fibre are used as fuel and for making gunpowder charcoal.

Justicia quinquangularis Fam. Acanthaceae. Annual summer season weed. In cotton, CMU herbicide per-emergence 1 lb/acre prevents emergence of the weed. Likewise aemazine 2-4 lbs/acre in maize and sugarcane is effective to control this weed.

K

Kainit A fertilizer source, mostly KCI, it is a good source of potash and has 20–30 per cent of K_2O.

Kani or levelling karah It consists of a flat blade of 3′ × 15″ with a vertical handle and two rings in the front for yoking the bullocks by means of ropes. This is probably the most common implement used in soil conversion and formation of bunds and levelling the land. Literally thousands of these are used in the Ahmednagar district of Maharashtra for bunding. Bunding of about five lakh acres has been accomplished by means of this one implement. Andhra Pradesh and Mysore, both use this for bunding of the land as well as for making nallahs.

Kaoliang A small grain of the millet group which is grown in N. China and Inner Mongolia. It can be used to produce flour or as animal feed. *See* also millet.

Kaolin (china clay) A fine white clay that easily crumbles to powder, resulting from the alteration of feldspars in granite. This alteration is caused either by weathering (hydrolysis) or by the action of ascending gases associated with igneous activity.

Two groups of this clay are commonly found (1) the kaolin, and (20 the montmorillonite. The molecules of the former are supposed to be composed of two sheets or plates, one of silica and one of alumina. Such clays are said to have a 1 to 1 type of crystal lattice. The two important representatives of the kaolin group are kaolinite and halloysite, do not exhibit colloidal properties.

Kaolinite The general formula is $Al_2O_3 2SiO_2 2H_2O$. Its crystal lattices are made up of one sheet of silica to one of alumina. It has a low base exchange capacity and little power to absorb water.

Karanda *Carissa carandus* Fam. Apocynaceae. A good plant for hedge and also a good jelly is made from ripe fruits.

Ker *Capparis deceoline*. Its fruits are very much valued as raw material for various pickles and grows naturally in the arid conditions of climate.

Kew barometer A mercury barometer that allows the atmosphere pressure to be read directly from the meniscus of a column of mercury contained in a graduated tube. *See* Barometer.

Khejri *Prosopis cineraria*. A dominant tree of the arid climate. It also yields fodder of quality. Immature pods are used as vegetables after cooking.

Khirni *Mimusops hexandra*. Fam. Sapotaceae.

Kind A species or sub-species of crop, plants known by one common name. For example: maize, paddy, and jowar in seed certification.

Kinds of phosphate fertilizers

1. Water soluble $[CaH_4(PO_4)_2$ $(NH_4H_2PO_4)]$ (K-phosphate) 2. Citrate soluble, in 15 per cent neutral ammonium citrate $Ca_2H_2(PO_4)_2$ In 2 per cent solution of citric acid-phosphates of basic slag.	Available phosphoric acid.
3. Insoluble phosphate of bone and raw rock phosphate	Unavailable phosphoric acid.

Kinds of soil alkali There are two kinds of alkali, white and black. White alkali consists of one or more of the above mentioned salts except sodium or potassium carbonate. Sodium and potassium carbonate, on the other hand, are called black alkali because they dissolve and disperse organic matter in the soil, giving it a dark brown or black colour.

Kinds of soil organism (1) Microflora: (*i*) Bacteria, (*ii*) Actinomycetes, (*iii*) Fungi, and (*iv*) Algae. (2) Microfauna: (*i*) Protozoa, and (*ii*) Nematodes. (3) The soil also harbours a large number of worms and insects of different kinds and sizes. Soil fauna, protozoa are the most abundant, followed by nematodes, worms and insects.

Kitchen garden Also called home-garden. It is small garden of vegetables or fruits taken in the small spare land of the house. It is very useful economically and not only gives substantial quantities of vegetables but is also helpful in keeping the environment clean and healthy by the plants that are grown around it.

Knots After the implements are used, they should be oiled so as to retain the polish of the steel and also to avoid rusting. The steel implements should be kept in a covered shade.

Kodo millet (*Paspalum scrobiculatum* I) It is hardy and remarkably drought-resistant food grain, grown on a fairly large scale in country of scarcity rainfall areas. The crop is usually grown on gravely, light upland poor soils.

Koppen climatic classification A climatic classification first proposed in 1918 by the German climatologist W. Koppen and subsequently modified several times. In the classification, the Earth is divided into temperature zones based on mean monthly temperatures. The four major zones are:

1. Tropical, all months over 20°C.
2. Subtropical, 4–11 months over 20°C and 1–8 months between 10°C and 20°C.
3. Temperate, 4–12 months between 10°C and 20°C.
4. Cold, 1–4 months between 10°C and 20°C and 8–11 months below 10°C.

The temperature zones are subdivided on the basis of seasonal incidence and amounts of annual rainfall.

Kumquat Fam. Rutaceae, *Fortunella japonica*. The fruits used are ornamental fruits. Oval shaped are *F. margarita* and the round shaped are *F. japonica*.

L

Labels An efficient lot numbering system which will maintain separate identity of each lot of seed is essential. A system of coded lot numbers can show any necessary information including crop, variety, field number, year grown and individual lot.

Labelling It is a part of a product which carries verbal information about the product or the seller. It may be part of packaging.

Labour intensive Increase in the employment through promotion of labour intensive for exportable agricultural commodities seen as an important strategy to promote equity.

Land management The legislation passed for the consolidation of holdings and prevention of fragmentation of agricultural holdings, promotion of cooperative farming and legislation in favour of agricultural labour form a part of the steps taken in the direction. Tenancy measures attempt to rationalise the land distribution system for progressive attainment of social equality and justice, it is also necessary to look on the problem of proper land use and management.

Land mortgage bank Loans can be obtained from a land mortgage bank. Obtaining loans from the central bank, it assigns the mortgage deeds obtained from the borrowers to the central bank. The central bank floats debentures and raises funds against the security of these properties. The State Government guarantees the repayment of the principal and interest on the debentures.

Land rent This comprises the net rental value of land owned by the cultivator and for the actual net rent paid for the land taken on rent.

Landscape The visual sum of all the landforms in an area together with the vegetation and the land use that is employed. Many efforts have been made to assess the quality of a landscape in a quantitative manner. These concern the use of landscape evaluation techniques, which attempt to assess a landscape with reference to its performing a particular function.

Land use classification The classification of land according to the use to which it is put. In the UK, the First Land Utilisation Survey, carried out in the 1930s by L.D. Stamp, identified six areas of land use–arable, barren and rough pasture, orchards and nurseries, meadowland, forest and woodland, and urban areas. The second survey to begun in 1960 and conducted by Alice Coleman, was based on 13 major classes of land use.

Lantana camera Fam. Verbenaceae. A perennial shrub weed. A mixture of 2, 4-D plus 2,4,5-T (brush killer) at 1–2 lbs. a.e./acre in 150 gallons of water at pre-bloom stage.

Late planting To sow a crop later than the right sowing time of the main crop. Generally early maturing varieties are preferred for this type of sowing.

Latin square (Sta) It may happen that fields chosen for laying out experiments sometimes exhibit fertility variations, therefore, depending on the uniformity of land chosen for experimentation as well as the shape and size of blocks and plots. These should be such as to result in the greatest homogeneity within blocks. This is possible in strips, that is, there may be alternate strips of high and low fertility caused by cultivation. A layout in randomized blocks will be effective in this situation if the blocks happen to be parallel to such strips, on the other hand, the layout will be extremely inefficient if the blocks to be placed across the strips. An ingenious method of eliminating such variation in fertility consists in an experimental arrangement, which simultaneously controls variation in two directions at right angles.

Law of diminishing returns An economic law stating that at a certain level of production, additional units of input will yield proportionately smaller units of output and the additional cost incurred will be greater than the additional revenue received. For example, the addition of more fertilizer to field of corn will at first increase the yield per hectare considerably, but a point will be reached at which the additional yield will have lower value than the cost of the additional fertilizer which is applied.

Lawns This is a mean of beautify a place by providing a green lively effect to a garden.

Lawsinia alba An important hedge plant *(Mehendi).*

Layering This method is striking cuttings that are still allowed to maintain their connection with the mother plant, by means of a portion, at least of their stem. Those plants in which propagation cannot be done by other methods, this method is successfully used in those cases.

Layout Before planting an orchard or seeding (sowing) of crops or starting a farm for growing crops or others, a mapping for the future requirement of growing the crop is needed. A proper layout always helps in future for spacing proper crops or trees and also for further cultivations to be done on the farm and orchard.

Leaf propagation Leaves are used as cutting. Roots come out either from margin of leaf of the surface or leaf blade.

Leafy vegetables Such plants of which only the leaves are used as green vegetables such as spinach, coriander etc.

Legislation for the preparation of compost Some Governments have made legislative provisions for the preparation of compost manure from municipal areas. According to this kind of manure Act, a committee is set which has the responsibility and powers to suggest to the Government to notify a particular area for the purpose of conserving the manure and making it incumbent on cultivators to take such measures, may be necessary for the purpose.

Legumes Among the green forage crops, legumes occupy a place of particular importance. Of all the common roughages, they are the richest in carotene, calcium and protein. The protein contained in them is of a high quality and would appear to supplement effectively the deficiencies of protein in cereal grains.

Legumes bacteria Certain plants such as clovers, peas, beans and other leguminous plants improve the soil in some way, making it possible to grow larger yields of cereals after these plants have occupied the land. This benefit has been traded to the fixation of nitrogen through the agency of bacteria contained in the nodules on the roots of the host. Such plants so affected belong to the family Legumoseae.

Legume vegetables Legumes form a group of crops belonging to the family Lelguminosease, which are grown not only for fruit and seed production but also for green fodder and silage, and as green manure. Their effect on soil fertility due to their deep penetrating roots, ability to fix nitrogen and value as soil cover is well known. Both the pods and seeds are rich in protein and are an excellent source of vitamins and minerals.

Peas is essentially a cold weather crop and can withstand light frost. Beans, of which there are several varieties, prefer a warm climate, have high water requirement and are sensitive to frost. Both are sensitive to high soil acidity. They are both of great value as an edible vegetable and for canning.

Legume vegetables sowing time In the hills, the main crop of peas is sown from April to end of May and a second crop in autumn. French beans are sown in the hills during May and June. In the plains, the main crop of peas is sown from September to November, and in some places a rainy season crop is sown in June and July. Beans are sown during February and March for an early crop, and in June and July as a late crop.

Legume vegetable seed rate Seed rate for peas varies from 40 to 60 lb., for French beans 50 to 60 lb., for cluster beans from 10 to 15 lb., for cowpeas from 20 to 25 lb. and for double beans from 12 to 15 lb. per acre.

Leguminous crops Crop plants belonging to family Leguminoceae, such as pulses, vegetables. More useful to grow them as inter-crops as they add to the grower and add fertility to soil.

Leguminous fodder crops The important leguminous fodder plants *Vigna sinensis* (cowpeas), *Cymopsis tetragonoloba* (cluster beans), *Phaseolus aconitifolius* (Moth beans), *Glycine max* (Soya beans),

Macuna spp. (velvet beans), *Trifolium alexandrium* (Egyptian clover), *Medicago sativa* (Lucerne or alfafa), *Melilotus perviflora* (Indian clover), *Trifolium resupinatum* (Persian clover), *Trigonelle foenum-groecum* (Methi) and others.

Lemma The other bract of the flower of grasses, sometimes referred to as the flowering glume.

Lemon (*Citrus lemon* Osback) It is not cultivated to any great extent in India as it requires a comparatively cool climate for regular bearing. The fruit is not so highly flavoured as sour lime.

Lentil (Lens culinaris Medic. Syn. *L. esculenta* Moench) Lentil is grown in large-areas. As pulse, lentil ranks equal to pigeon pea and inferior to black gram. Unripe pods are used as green vegetable, and dry leaves, stalks, husk and broken grain as cattle feed.

Leucas aspera Fam. Labiateae. An annual weed. Susceptible to 2,4-D amine 1 lb a.e. in 100 gallons of water spray.

Level An instrument used in surveying to determine height. It consists of a spirit level mounted on a telescope, which swings in a horizontal plane. It is used in conjunction with graduated levelling staves. *See* Abney level, dumpy level.

Levellers and clot crushers (The indigenous implement is called *patela* or *sohaga*.) It is merely a flat log of wood 6′ to 8′ in length pulled by bullocks over the clods. The driver usually stands on the log to give it weight. In Uttar Pradesh, improved *pata* or *patela* is now mass manufactured and has become quite popular. It has a series of pointed hooks made of ¼″ or 5/8″ mild steel rod. These hooks help in crushing the clods and in collecting weeds.

A land leveller or *jandra* is also used in some-facts. It consists of a plank about 3′ to 4′ long, 5″ wide and 1″ thick. In those areas where bamboo is readily available, a ladder-shaped leveller made of bamboo is used.

Ley farming Ley farming consists of growing grass in rotation with agricultural crops. In England and America, this system has been extensively used to improve the fertility of the soil and provide good fodder to the animals. Ley farming also helps in building up the structure of the soil and preventing soil erosion.

In India, this method can be conveniently adopted in Nilgris and other similar places which abound in good pastureland and have suitable climate.

Life of papaya plants Most strains of papaya tend to grow as single stems until the plant is several years old, when some branching may occur. But this practice is generally discouraged for growing a commercial crop.

Lifting water appliances Worked by Human or Bullock Power. The most commonly used water lifts of this type are *picota, mhote* and Persian wheel or *rahat*. The picota is worked by human labour while mhote and Persian wheel are bullock operated.

Lime (*Citrus aurantifolia* Swingle) The lime, both sour and sweet, known as *Kaghzi nimboo* and *mitha* respectively, are more prized in India than lemon. Sour lime is propagated mainly from seed. Budding on rough lemon rootstock, layering and maorcotting are also practised to some extent. The tree is susceptible to frost. It flowers twice a year in February–March and again in August. The main crop is obtained in August from the first flowering. The second crop is ready in the following February.

Propagation of sweet lime is done from mature woodcuttings which root readily. It can also be propagated from seeds and the seedlings usually come true to type. Planting of limes and their pruning, manuring etc. are the same as for sweet orange.

Lime and green chilli pickle Fully ripe limes are washed and cut into halves or quarters. The stalks of chillies are removed and a slit is given. For every pound of lime and green chilli used, 4 oz. of powdered salt is added, and put into a clean sterilized wide-mouth glass or glazed jar. The limes are partially squeezed. If the juice is not sufficient to cover the enteric mass, a solution of citric acid (4 to 5 per cent) is added. The entire mass is kept in the sun for about a week when it becomes ready for use.

Limiting factors When a process is conditioned, as to its rapidity by a number of separate factors, the rate of the process is limited by the pace of the slowest factor. In general, the behaviour of a system depends on a number of different factors; under certain conditions, one of these can limit the behaviour.

Lining of irrigation channels Economic and efficient channel lining goes a long way in conserving the water. Brick lining is widely used. Stone, wood and tiles are also used sparingly. Cheap lining of bituminous-straw mixture is receiving a good attention, tiles and polythene etc are used for lining of channels.

Linseed (*Linum usitatissimum* L) The oil produced from the seed of the flax plant. The oil is used in paints, varnishes and linoleum, and residue is compressed into a cake, a cattle feed. *See* also flax.

Linseed has been grown from very remote times for the manufacture of linen and for its seed which contains a large quantity of oil. The oldest regions of cultivated flax are reported to be in Asia and on the coasts of Mediterranean.

Linseed is grown mainly for seed used for oil extraction. The oil-content of the seed varies from 33 to 47 per cent in different varieties. Linseed oil is an excellent drying oil which is extensively used in the preparation of paints, varnishes, printing, inks, oil cloth and waterproof fabrics, and as edible oil in some areas. The residue oil cake is a valuable cattle feed as well as manure. The stalks are used for fuel, but sometimes coarse fibre is extracted from them.

Linseed cake It has 5.5 per cent nitrogen, 1.4 per cent phosphorus and 1.3 per cent potash.

Liquid manures Any manure or fertilizer given in the liquid form, generally given as mixed in water. In some foreign countries, liquid manures are given to various crops.

Litchi (*Litchi chinensis* Sown) Fam. Sapindaceae. Litchi is grown extensively in north Bihar, sub-montane districts of western Uttar Pradesh and the Punjab. Hot dry winds in summer when the fruit is ripening are very undesirable. It grows well in sandy loams and clay loams that contain large quantities of lime.

Litre A metric unit of volume defined as 10 metre or 1000 cm.

Litter The vegetable matter mainly fallen leaves and twigs that loosely covers the surface of soil, sometimes to a depth of many centimetres.

Little millet (*Panicum miliare* Lamk) It is grown on a limited scale as a poor man's crop, capable of withstanding both drought and water logging. The method of cultivation is similar to that of barnyard millet so also the yield and uses of grain and straw.

Livestock The fact that all livestock work is based on utilitarian principles makes it necessary for an animal to be provided with the environmental conditions most conducive to its full inheritance being expressed in production. It has to be fed with adequate quantities of a balanced ration and kept free from disease. It should be managed, house and looked after during growth, pregnancy, lactation and dry period.

Local climate The climate of a small area that is distinct from the climates of surrounding areas.

Local control (Sta) The random allocation of treatments to plots while giving an estimate of treatment difference freed from any systematic influence of environment or bias as well as providing a correct test of significance of the difference is not quite efficient. It would be obviously desirable to reduce the experimental error as far as practicable without interfering with the statistical requirement of randomness, because a lower experimental error means that a smaller real difference between treatments can be detected to be significant.

Local or desi ploughs This is the primary implement used all over world. Desi ploughs have almost a common design and they differ mostly in the size, depending upon the type of soil and the size of the bullocks and animal power used in the particular areas. The heaviest ploughs are found in the black cotton soil areas of India. The desi plough is almost like a western style cultivator. The body and the wedge may be of one price or two-piece construction. The iron point or share is made up of a mild steel rod or bar and helps penetration into the soil and protects the wooden wedge against wear and tear. The beam varies from 9 to 11 in length and is made up of the body. All indigenous ploughs are similar in design, but vary considerably in size and weight to suit the prevailing soil types, ploughing conditions and the weight and height of draft animals in the locality. The working draft varies from 60 to 350 Ibs.

Localised placement By this is meant to place the manure near the seed or plant. This is advantageous when manure is available in a small quantity.

Lona phosphate It contains 18–20 per cent nitrogen and 49 per cent phosphorus.

Logan-berry *Rubus longbaccus*. Fam. Rosaceae.

Longan *Euphoria longana*, Fam. Sapindaceae.

Loss of biological diversity One of the most controversial aspect of agriculture research has been the loss of biological diversity due in part to the introduction of improved varieties of plants and animals. It has been shown that the introduction of such varieties displaces the traditional ones and therefore undermines the very genetic base on which modem agriculture rests.

Loppers These are shears used during pruning of branches at greater heights. Most loppers are intended to prune branches not of more than one inch diameter. Use of loppers greatly increases the speed of pruning and if they are kept sharp and carefully used, satisfactory work is done.

Loquat (*Eriobotrya japonica* Lindl.) Fam. Rosaceae. Loquat is grown mainly in the Punjab and Uttar Pradesh, about 800 hectares each, and to a small extent in Delhi, Assam, Maharashtra and the hills of south India. While the tree can stand temperature well below freezing, the flowers and young fruits which are borne in winter are severely injured by frost. It does not fruit well in areas having warm winter. It thrives on many different types of soils.

Lot numbers Every seed processing plant should have a system of identifying seed lots by lot numbers. Each lot should be assigned a number and each container of seed should be labelled with lot numbers.

Low land An area of land of low elevation. It may be low relative to adjacent lands with a higher elevation (*e.g.* the Scottish Lowlands) or it may be an area of land that lies close to sea level (*e.g.* the Fens).

Low land cultivation In low land areas where water keeps accumulated for long time, long duration photosensitive crops are grown especially rice. Here the depth varies from 15 to 100 cms and more.

Lucerne *See* alfalfa.

M

Magnesium deficiency It is shown by plants as their leaves dry up but the midrib remains green. In cotton, this colour becomes violet-red but the veins remain green. Gradually the leaves finally turn brown and start dying.

Magnetic separator Magnetic separators of seed processing machines were originally designed for industrial uses such as ore separation. Later an increasing number of applications in seed processing have been found for magnetic separators.

Mahuwa cake A rich organic fertilizer. *See* cakes.

Maintaining seed lot identity Complete identity of every lot and bag of seed at all times is essential. Since both the conditions and the containers of seed lots are changed during processing, adequate records and proper labelling during process are necessary.

Maize (Indian corn) A grass cereal that produces large edible seeds. It is known as corn in the USA. Maize requires a growing season of at least 140 days and 200 mm of rain in the summer months. The plant grows to a height of 3-4m and produces seed cobs on the main stem. It forms a staple food in South America and many other parts of the world. From the seeds, a number of food products, such as corn flakes, popcorn, corn oil, corn beer and cornflour, are obtained.

Major grass covers in India

Sehima-Dichanthium

Dichanthium-Cenchrus-Elyonurus

Phragmites-Saccharum

Themeda-Arundinella

Temperate and alpine grasslands

Mallee A scrub vegetation that grows in parts of SW and SE Australia under desert conditions. Mallee consists mainly of ever green eucalyptus shrubs, about 2–3m high, forming dense low thickets.

Management practices for weed control Such as proper and timely fertilisation and irrigation serve a dual purpose as weed control measures as well as their direct stimulatory effect upon the potential yield of crops. Many weeds thrive under low fertility conditions while crop plants cannot. Increased fertility shifts the competitive balance in favour of the crop.

Mandarins These are citrus fruits like *Nagpuri Santra* (orange) and are grouped containing similar varieties such as Coorg orange, and Assam oranges or oranges grown mostly on hills in general. These are loose skinned oranges.

Mandarin orange (*Citrus reticulata* Blanco) Santra or mandarin orange grows successfully in all tropical and sub-tropical parts of the country. It tolerates more humidity in summer and winter than the sweet orange. It is grown under rainfed and irrigated conditions. In the Punjab, its cultivation is confined mainly to the sub-montane districts up to about 600 metres. It can be grown successfully on a wide range of soils, but the ideal soil is medium or light loam with a slightly heavier subsoil. Heavy black soil, underlaid with murram and having good drainage, is also suitable. In the Khasi Hills of Assam, oranges are grown on sandy or gravely soils.

Manganese Plants take it in the form of Mn++ and Mn+++ form. It helps in the formation of chlorophyll. The deficiency of this in plants is visible by the presence of dead fibres and tissue on the leaves in an irregular manner. The small veins remain green. When the pH of a mineral soil is low, appreciable amounts of manganese along with iron arid aluminium are active, so much so in fact that the aluminium may become extremely toxic to certain plants.

Manganese deficiency *See* witchis brooms in peach and apricot.

Mango (*Mangifera indica.* L.) Fam. Anacardiaceae. Mango occupies nearly half the total area under fruits in the country. It is adaptable to a wide range of soil and climatic condition, and grows well right from Assam to the southern-most limit of the country and from sea level upto about 1,500 metres. It withstands both fairly dry and heavy rainfall conditions, provided severe and recurring frosts in winter do not endanger the young tree.

Mango chutney (sweet) Fully mature mangoes, that are not soft, are washed, peeled and cut into slices. The slices may be directly used for making chutney, or stored in 15 per cent brine. If brine is used, the salt solution should be leached from the slices before use. For every 2 lb. of mango slice, 2 lb. of sugar, 2 oz. salt, 1 oz. mixed spices (equal quantities of cardamom, cinnamon, cumin, etc.) 1/5 oz. of chopped garlic, ½ oz. of finely ground red chillies, 4 oz. of vinegar, 1 oz. of chopped onions and 4 oz. of green ginger are used in the preparation of chutney. The slices are heated in a small quantity of water to make them soft, and salt and sugar added. The rest of the ingredients tied loosely in the cloth bag are added and the slices are cooked to the consistency of a jam. Vinegar is now added, and boiling continued for another five minutes. The bag of spices is removed and the product is cooled and bottled.

Mango pickle Fully developed under-ripe mangoes are washed in water, cut into slices, stones removed and placed in brine to prevent blackening. Two pounds of mango slices are mixed with 8 oz. of common salt in a glazed jar and kept in the sun for three to five days. Four ounces of coarsely ground fenugreek, 1 oz. of coarsely ground nigella (Kalaunji) and 1 oz. each of turmeric powder, black pepper and aniseed are mixed with cured mango slices smeared with a little rapeseed (sarson) oil, and the mixture packed tightly in glass or glazed jars, and covered with a thin layer of oil and sealed. The pickle will be ready for bottling in two to three weeks.

Mango showers Heavy showers, often accompanied by thunder, that occur in India between March and May, before the breaking of the main monsoon. The name arises from the fact that at this time the mangoes are beginning to ripen. *See* also blossom showers.

Manila hemp *Musa textilis*. Fam. Musaceae. An industrially important plant of Philippines.

Manual labour expenditure This item includes the wages of hired workers actually paid out, whether permanent or casual as well as the estimated cost of family workers.

Manufacturing industry Those industries that process materials or assemble components to produce finished products. Manufacturing industry is distinct from other types such as service industries or the tourist industry because goods and materials are produced.

Manuring There are certain rules which must be followed in the supply of manures to the plant and trees (1) Manures should be given so that they do not harm the roots. (2) Manures which are suitable for being given to special kinds of soils and trees should be selected one. (3) Manures given at right time in right quantity is always beneficial for soil as well as plants. (4) While planting new trees, the pits should not be filled with manures. (5) Those manures which are easily soluble in water should not be given before rainy season starts because they are likely to go deep in the soil with rain water. (6) Such fertilizer mixtures should not be used which may damage the plants and soil. (7) While using chemical manures, it should be taken care that these do not fall on the leaves and buds of the plants. (8) Both organic and inorganic fertilizers should be equally used as they are complementary to each other. (9) The new and young saplings should not be fertilized at the time of planting. (10) Fertilizers should be seldom used than frequently. (11) More organic manures should be given to potted plants. (12) The use of liquid manure is done at different times in different plants *e.g.* for flowering plants at the time of flower bud formation; on using on fruit tree the colour of fruit improves, vegetable should get manures at the time of their growth and in pots when the plants have fully developed their roots. (13) If there is scarcity of irrigation water then green manuring should not be done.

(14) Fertilizer should be given in such a way that it does not touch stem of the trees.

Manurial trials To test various doses of fertilizers and manures for the optimum economic yields in view of the cost of fertilizer suited for this purpose. The main consideration of this kind of test is only economical.

Marcottage *See* air layering. Also called Gootee.

Margin of production The point at which the cost of production is just covered by the profits obtained from the sale of the produce.

Marmalades It is citrus fruit jelly containing shreds of the fruit used. Malta or sathgudi and Khatta (sour orange) or galgal are used in the ratio of 2:1. Peel the outer layer of the skin very thinly. Cut the peeled fruit into thin slices, cover with two to three times its weight of water, and boil for about 30 minutes. Press out the juice, add 14 to 16 oz. of sugar per lb. of extract and boil. Cut some of the peels into thin slices (3/4 to 1) and boil them in water till soft and drain. When the temperature of the juice extract reaches about 218°F, add the shreds and boil further till the temperature reaches 221°F. Finish boiling as in guava jelly. Cool the product for a few minutes and pour into clean dry jars to set and then seal.

Mass production The manufacture of a commodity on a very large scale. This is normally achieved in a factory by conveyor belt techniques or the continuous use of automatic equipment. The high volume of output provides economies of such scale, which may cheapen the commodity. For example, mass production techniques have reduced the comparative prices of TV sets and cars in recent years.

Mass selection A system of breeding in which seed from individuals selected on the basis of phenotype is composited and used to grow the next generation.

Material index The measurement of the loss of weight during the processing of raw materials when compared with the weight of the finished product. The formula is:

$$\text{Material index} = \frac{\text{Weight of raw material}}{\text{Weight of finished product}}$$

The higher the index is above 1 the greater the significance of the cost of moving the raw material to the factory.

Maturity of tests of citrus The sweet species are not harvested until reasonably mature, and in some countries law forbid their sale until they have attained a certain proportion of total solids to acid. 8 Parts of total solids to one part acid is considered most satisfactory.

Maximum thermometer A thermometer that records the maximum temperature achieved over a specified period of time. The most common type consists of an ordinary mercury column. The marker can be pushed upwards by the mercury but cannot plied down again as the temperature drops and the mercury recedes. *See* also minimum temperature.

MCPA A herbicide. 2-methyl-4-chlophenoxy acetic acid.

Mean (Arithmetic mean) In statistics, the average value. It is obtained by the formula:

$$X = \frac{SX}{n}$$

where, X is the mean, *n* is the number of values being considered and SX is the sum of all the values. *See* also median, mode.

Mean (Sta) It is the arithmetic average and is the result obtained when the sum of the values of each individual in the data is divided by the number of individuals in the data. The mean is usually denoted by the symbol *u* and given by $u = \frac{\Sigma X}{N}$, *where* Σ represents the summation which is taken over all the observations in the population. When the data are expressed in a frequency distribution, the mean is calculated from the formula:

$u = \frac{\Sigma f.x}{N}$, *where* f is class frequency, *x* is the population and Σ indicates summation of the N products f with x over all classes.

Mean annual rainfall The distribution of the mean annual evaporation in any country is important. Some area of highest mean annual evaporation as one moves away from this area northwards, eastwards or southwards, the annual rate of evaporation is proportional to the wind velocity and the dryness of the air.

Meander A pronounced curve or loop in the course of a river channel. The flow of water in a meandering stretch of river is not even. The current is fastest on the outside of the meander bend and this area is subjected to the greatest erosion. Concentrated erosion on the outside of meander bends results in the formation of a narrow neck. The river may eventually break through the neck in times of flood bypassing the meander bend and forming a cut-off, the ends of which rapidly silt up forming an oxbow lake.

Meander belt The part of a valley floor over which a river will shift its channel. The boundaries of a meander belt are marked by the concave banks of successive meanders. Beyond the meander belt, there may be low river bluffs or river terraces.

Meandering valley A sinous valley with pronounced curves. The slopes on inside of these curves tend to be less steep and frequently convex, whereas the slopes on the outside of the bends are steeper and may be concave in shape. The river that occupies the valley is likely to have developed meanders on a much smaller scale over the alluvial deposits of the valley floor.

Mean deviation (Sta) Another measure of dispersion is provided by the mean deviation from their arithmetic mean without regard to the sign and dividing their sum by the number of observations. The formula for calculating the mean deviation for data classified in the frequency distribution is:

$$\text{Mean deviation} = \frac{\Sigma f.d}{N}$$

where, f is class frequency, d is deviation of the mid-value of the class from the population mean, always taken positive, N is total number of observation and the summation Σ is taken over all the classes of the distribution.

Mean sea level The average level of the surface of the sea It is calculated from data which takes into account the changes in tide that have occurred over a long period of time. In the UK, this calculation is based on Newlyin in Cornwall (the Ordnance Datum).

Measure of location (Sta) A single member which enables us to assess the position in which a group is located with respect to other groups. The mean, median and mode are the examples of measures of locations.

Measurement of irrigation water Irrigation water is measured in the following units, *viz.* (*i*) cusec, (*ii*) acre-inch or acre-foot, and (*iii*) gallons per hour. Cusec is the quantity of water flowing at the rate of one cubic foot per second. As one cubic foot of water weights 62.4 lb., 1 cusec flowing for one hour is equal to 62.4 × 60 lb. or approximately 100 tons.

Measurement of moisture in soil A knowledge of the amount of moisture present in a soil is very necessary for determining the time and amount of irrigation. A practical method to measure soil moisture quickly and accurately is very useful to a farmer to enable him to find out when his crop requires to be irrigated.

Measure-weight test The weighing of a specific volume of seed gives an indication of its relative heaviness, plumpness and quality. Measure weight devices give kilo per hecto-litre.

Mechanical methods of weed control Hand pulling or hand weeding, hoeing, tillage, mowing, burning, flooding, smothering etc. are examples of mechanical methods. Hand weeding is the most efficient method but it is back-breaking, time-consuming and costly. Tillage is a practical and economical method of controlling annual weeds. The plough (disc, mouldboard, etc.), the harrow (disc, spike and spring tooth) and the cultivator (duck-foot, blade, bakhar etc.) are some implements in vogue.

Mechanism of nodule fixation Entrance of the organisms is affected through root-hairs. Through the entire length of root hair, the tube further extends into the cortex of the root. Here growth of tubercle starts and fixation of nitrogen also takes place. It seems that the nitrogen compounds produced within the bacterial cells are diffused through the cell wall and absorbed by the host plant.

Mechanism of phosphate release In flooded soils, it has been found that this is released (1) reduction of insoluble phosphate to a more soluble form (2) release of occluded phosphate by reduction of hydrated ferric oxide coatings (3) displacement of phosphate from ferric and aluminium phosphate by organic anions (4) hydrolysis of ferric and aluminium phosphates due to an increase in alkalinity and (5) anion exchange between clay and organic anions. Certain liberate hydrogen sulphide which reacts with ferric phosphate to form ferric sulphide thus liberating the phosphate.

Median In statistics, the central value or class in a series of ranked values. *See* also mean, mode.

Median Sta The median is the value which is located in the middle of a series when the observations are arranged in order of magnitude and it divides the series into two halves, half the number of observations in the series, thus if 101 observations are placed in the order of their magnitude, the 51st observation will be the value of the median.

Medicago hispida A good fodder legume plant.

Medicinal value Of fruits. Fruits are not medicine but it is recognised that as they provide proteins, fats and carbohydrates, vitamins, minerals, some enzymes etc. their use itself implies that they should be eaten so that medicines, are not necessary to use. Fruits have a great dietic value.

Mediterranean climate (warm temperate western margin climate) The broad climatic type that occurs on the W coasts of the continents between latitudes 30° and 40°N and S. The major influence in these areas is the seasonal shift of the wind belts, with the trade winds blowing off the land in summer and the westerlies blowing off the sea in winter. Thus summers tend to be hot and dry while in winter, depressions move in from the west bringing mild moist conditions.

Mega-symbol M A prefix denofing 10^4, for example, 1 megahertz (MH_2) = 10^6 hertz.

Melilotus alba It is good fodder legume plant. An annual weed. Can be controlled by 2,4-D in cereals at 1 lb. a.e./acre, kills the annual legume. For the perennials a mixture of 2,4-D and 2,4,5-T at ½ lb. a.e./acre prior to flowering.

Melilotus indica Same as *M. alba*.

Mesosphere A layer in the atmosphere extending between the atmosphere (at an altitude of about 50 kms) and the mesosphere (at about 80–90 kms). Within the atmosphere, the temperature decreases with altitude from about –100°C at the mesopause.

Mesta (*Hibiscus cannabinus* L.) Mesta, Decan hemps or Bimlipatam jute is grown in many parts of India. The crop is grown for fibre, which is used for cordage, fishing nets and small quantities of rough sack-cloth and canvas. It is also mixed with jute in the manufacture of sugar bags. Leaves, tender shoots and fruits are sour and are consumed as pot herb. Tops and branches are fed to cattle. Seed is used as feed for milch cattle, dried stalks as fuel, and dried, split stalks for match sticks.

Meta-phos It is fertilizer not commonly used. It is a calcium metaphosphate $Ca(PO_3)_2$, has 62.63 per cent available P_2O_5.

Meteorological service Meteorological departments are served by a network of meteorological observations all over the country. There are surface observatories, pilot balloon stations, radio-sound stations, hydro-meteorological stations, crop weather observatories and about rain-gauge stations. They collect wind, rainfall clouds, crops etc. on the basis of which daily weather reports, forecasts and warnings are prepared by Regional Forecasting Centres.

Meteorology The scientific study of the atmosphere and the physical processes at work within it, including pressure, winds, temperature, precipitation, clouds, sunshine etc. The world-wide collection and study of data relating to these processes over a long period of time has enabled meteorologists to discern recurring patterns of weather development, which are the basis of weather forecasting. The forecaster uses a weather (or synoptic) chart plotted from data obtained simultaneously at a great number of weather stations and from this it is able to predict the likely weather conditions for a limited time ahead.

Method of applying lime It is best applied to ploughed land and worked into the soil as the seedbed is prepared. It is thoroughly mixed with the surface half of the furrow slice. The soil should not be too moist when application is made.

Methods of composting There are two methods of composting: (1) Aerobic, and (2) Anaerobic methods.

Methods of irrigating fields There are four different methods and three principal methods of irrigation, *viz*. Surface, sub-surface and aerial, overhead or sprinkler irrigation.

Methods of study of percolation and leaching Two general methods of procedures are available for study of percolation and leaching (1) the use of an effective system of tile drains and (2) construction of lysimeter.

Methods of weed control These are divided into mechanism (cultivation and mowing), cultural or cropping, biological and chemical. Each of these or a combination of these to secure efficient weed control at economical cost.

Micro-organisms These are studied to test the presence of quantity of the nutrients in the soil. Especially such micro-organisms as *Azotobacter* and *Aspergillus niger* are used for this purpose. Two solutions are prepared from the soil-one is neutral and the other normal. The micro-organisms are counted in both after keeping them for some time.

Micro-organisms in soil Raw organic matter in the soil is not directly used by plants as food. It must be broken down first into humus and then into simpler products before it can be so utilised. This work is done by different kinds of micro-organisms which inhabit the soil by the billions. Decompositions of organic matters primarily forms part of the feeding and growth processes of these microscopic plants and animals. Sugars, starches and proteins are broken down first; then cellulose and fatty substances (lipoids); and lastly lignin or woody substances. When organic matter becomes chemically and biologically inherit, the soil becomes infertile despite its presence. However, not all the soil organisms are beneficial. There are certain bacteria which under anaerobic conditions of waterlogged soils cause denitrification, releasing nitrogen which gets lost into the air. Some others cause plant disease.

Micronutrient A nutrient required in trace amounts by an organism. For example, a plant can obtain sufficient of essential trace element manganese from solution containing 0.5 parts per million of manganese. Micronutrients include trace elements and vitamins. *See* deficiency. There are certain elements which plants need in very small quantities for their proper growth and functions. These are Iron, Manganese, Boron, Molybdenum, Copper, Zinc and Chlorine. These seven elements are called rare elements or minor elements or trace elements.

Mid interval values This is used to calculate the mean weight of samples say, the mean weight of potatoes per root from a given date.

Mid-latitude Denoting those areas of the world lying between latitudes 30° and 60°N and S. In meteorological and climatic terms, it is the area within which the major factor is the meeting of polar and tropical air masses producing frontal activity.

Migration An instinctive regular two way movement of part or all of an animal population to and from a given area, usually along well-defined routes. It is closely linked to the cycle of the seasons and is triggered off by seasonal factors such as increasing and decreasing day lengths in spring and autumn. Many birds, hoofed mammals, bats, whales, fish and insects migrate, often covering immense distances. For example, the Arctic tern breeds on the northernmost coasts of Eurasia and America and winter around the Antarctic pack ice 11000 miles to the south. Migratory mammals such as the wildebeest live in habitats with fluctuating climatic conditions and migrate in order to find an adequate food supply.

Millets These are small grain crops grown on below marginal areas in poor soils. These are *Eleusine coracana* (Ragi), *Setaria italica* (Kakun), *Panicum milliceum* (Sawan), etc. *Bajra* (*Pennisetum typhoidem*) is categorised as major millet. This group of food grains include Barnyard millet, Bulurush millet, common millet, Finger millet, Italian millet, Kodo millet, little millet, Maize, Sorghum etc. Thse are the cereal grasses. There are many varieties of millet grown in Asia, N and E Africa and S Europe. The plant grows to cover 1 m in height and produces a head containing large numbers of small seeds. These can be made into flour for human consumption or used as a fodder-crop.

Millibar symbol mb The unit of pressure used in meteorology. It is equal to one thousandth of a bar *i.e.* 100 pascals.

Milli-symbol m A prefix denoting 10^{-3}.

Milling and *gur* making from sugarcane Three-roller iron crushers operated by bullocks, oil engines or electric motors extract 60 to

70 per cent of juice from cane and are in common use. Milling should be done in the cool hours of the morning and the extracted juice should be boiled without delay. Metal vessels are preferable as containers for juice. Begasse is almost the universally used fuel. But this is often supplemented with sticks of cotton, fur or castor. For economy in fuel and efficiency in boiling, improved furnaces such as Lucknow bel, improved.

Millioncity A very large city with a population of more than one million.

Mimosa pudica Fam. Leguminosae. A thorny perennial weed. Control same as for *M. alba.*

Mineralization of organic nitrogen In an anaerobic soil a sequence as: proteins polypeptides-amino acids-ammonia-NO_2-NO_3. The degradation of ammonia stages is bought about by general type heterotropie-micro-organisms like bacteria, fungi and actinomyces, while the subsequent process of nitrification is carried out by two specialised groups of aerobic bacteria. The first group, nitrosomonas, oxides ammonia to nitric and the second nitrobacter, oxidises nitrite to nitrate, thus:

Ammonia → Nitrosomonas → Nitriten → Nitrobacter → Nitrate

Minimum thermometer A thermometer that records the minimum temperature achieved over a specified period of time. *See* maximum thermometer.

Minor elements *See* micro-nutrients.

MH-maleic hydrazide 1, 2-dihydropyradazine-3, 6-dione, a herbicide.

Mhote It is an indigenous contrivance in which a large leather or iron bucket is hauled up and down the well by bullocks walking on a sioping ramp constructed near the well The bucket is pulled by ropes, one end of which is tied to the bucket and the other to the bullock's yoke. The ropes moves over wooden pulleys suitably fixed on the well. Generally one pair of bullocks is sufficient for drawing water from a depth of 20 to 30 feet. For deep wells, two or even three pairs are necessary. The bucket holds 40 to 50 gallons of water and the quantity of water lifted varies from 1,000 to 1,5000 gallons per hour over a lift of 30 feet. In one of the improved types of mhote, the bullocks walk on level land in a circle instead of on the ramp. The bullock power is transmitted through a wooden shaft to a rotating drum on which the rope carrying the bucket winds and unwinds as the bucket is raised or lowered.

Miscellaneous manures This name is given to such manures which supply two or more nutrient elements.

Mistletoe A flowering plant which is parasitic on fruit trees. These belong to genus Loranthus. They have thick succulent light green leaves pale flowers and bright berries. It has about 30 hosts among fruit trees.

Mixed cropping Farming that includes the growing of arable crops and the keeping of livestock. In many cases, part of arable land is used for growing fodder crops for the livestock. In the corn-growing area of the US Midwest, cattle and pigs are-reared and fed on part of the corn crop. Cash crops such as malting barley and potatoes are grown on mixed farms in the UK and W. Europe.

Mixed fertilizers Compound fertilizers contain plant food elements in fixed proportions and are, therefore, not always best adapted to different kinds of soils. Accordingly, the needs of different soils can generally be met most economically by the use of fertilizer mixtures containing two or more materials in suitable proportions. Mixtures usually meet nutrient deficiencies in a more balanced manner and require less labour to apply than straight fertilizers used separately. Mixtures containing all the three principal nutrients (N, P and K) are termed complete fertilizers.

Mixed flow pump This type has a fairly large capacity and can be installed on wells having casing too small for the centrifugal type. Turbine pumps with submersible motors are also available. All turbine pumps give long and efficient service when used under proper conditions.

Mixed inflorescence These occur in some varieties of mangoes, while only once a year such as Fazli. Such varieties while bear the fewer mixed panicles bear the heaviest crops and are the most markedly alternate bearers.

Mixing and dividing seed samples Preparation of seed sample includes the process as (1) mixing, (2) dividing, (3) moisture testing, and (4) determining of test weight. For all these processes, there are equipments to use and get quick work done.

Mixing of manures Sometimes for economical reasons, the fertilizers are mixed and given to the soil. But is not advisable to mix any fertilizer with another as it may immediately react chemically and be of no use to the soil and plants, and may have detrimental effect on the soil and plants. The following scheme as shown in the diagram should be followed before any scheme of mixing of available fertilizers and manures.

●	●	■	●	●	●	▲	■	■	●	●
●	●	■	■	●	●	■	■	■	●	●
●	●	●	■	▲	■	■	■	●	●	●
●	●	●	▲	▲	■	▲	●	●	■	■
▲	▲	▲	▲	▲	▲	▲	▲	▲	▲	▲
▲	●	■	■	●	●	▲	■	■	●	●
▲	▲	▲	▲	●	●	▲	▲	●	●	●
▲	▲	●	●	●	■	▲	▲	▲	■	●
●	●	●	●	▲	■	▲	●	●	■	■
●	●	●	▲	▲	●	▲	●	●	●	●
●	●	●	▲	▲	▲	▲	●	●	●	●
11	10	9	8	7	6	5	4	3	2	1

● **Fertilizers can be mixed.**

▲ **Can be mixed just before use.**

■ **Should not be mixed.**

1: Muriate of potash; 2: Potassium sulphate; 3: Ammonium sulphate; 4: Calcium ammonia nitrate; 5: Sodium nitrate; 6: Calcium cynide; 7: Urea; 8: Superphosphate; 9: Ammonium phosphate; 10: Basic slag; 11: Calcium carbonate

Mixture Two or more substances forming a system in which there is no chemical bonding between the two. In homogeneous mixtures (*e.g.* solutions or mixtures or gases), the molecules of the substances are mixed and there is only one phase. Heterogeneous mixtures' (*e.g.* gunpowder or certain alloys) different phases can be distinguished mixtures differ from chemical compounds in that.

Mixture No. 1 Equal quantities of ammonium sulphate, urea, superphosphate and bone-meal are mixed in this mixture.

Mixture No. 2 This mixture is made by mixing ammonium sulphate and superphosphate. In this, 30 parts of ammonium sulphate and 40 parts of superphosphate are mixed to provide 12 per cent nitrogen and 6 per cent phosphoric acid in this.

Mixture No. 3 This mixture contains ammonium sulphate, superphosphate and potash in it. This can provide 12 per cent nitrogen, 6 per cent phosphoric acid and 6 per cent potash.

Mixture of fertilizers The preparation of mixed fertilizers requires a good knowledge of the properties and mutual reaction of the component fertilizers under different climatic and storage conditions. Therefore, in preparing such mixtures at the farm or home, care should be taken to avoid uneven mixing of some of the fertilising nutrients in the form of gas, converts soluble nutrients into soluble ones or produces caking. It is unwise to mix:

1. Ammonium sulphate, ammonium chloride, other ammonical fertilizers and nitrogenous organic manures with lime;
2. Sodium nitrates or potassium nitrate with superphosphate;
3. Nitrochalk with superphosphate or lime;
4. Ammonium sulphate-nitrate with lime;
5. Urea with superphosphate;
6. Superphosphate with lime or calcium carbonate or wood ashes.

Mobility of elements in plants Nutrients are divided in two classes according to their mobility–(1) Nitrogen, phosphorus and potash, magnesium and zinc which have a greater mobility and the signs of their deficiency appear on the lower and old leaves of the plant, and (2) Calcium, boron, copper, manganese, sulphur and iron have lesser mobility. The symptoms of their deficiency can be observed on tender leaves, shoots and new leaves.

Mode of application of fertilizers and manures Bulky organic manures should be applied well ahead of sowing so that the preliminary decomposition takes place before the seeds germinate. Failing pre-sowing application, they may be applied any time after the seedling have established themselves. They are best applied in powdered form. A sufficient supply of moisture in the soil is essential for their rapid decomposition. In the case of inorganic fertilizers, potassic and phosphatic fertilizers are best applied just before sowing or transplanting. Nitrogenous fertilizers may be applied either at planting and partly later. Split application is particularly desirable for nitrogen when applied to irrigated crops or to crops in heavy rainfall areas. In advanced countries, fertilizers are usually applied with the help of machinery of diverse kinds and sometimes by airplane or a helicopter. Combined planters and fertilizers' distributors are employed when row crops are fertilized at sowing time.

Mode (Sta) The mode is that value of the variate which occurs most frequently. In a frequency table the modal class which has the greatest frequency. This class can be determined at once from inspection, but the actual value of the mode will be located somewhere in that class interval, not necessarily at the mid-point of the class.

Moder A form of humus that is intermediate between acidic mor and neutral mull. Moder shows some mixing with the coarser mineral particles in the soil, but retains a stratified appearance. Compare mor, mull. *See* humus.

Modified leader system The young tree is pruned as to be trained to give a final shape of a *vase* and is further pruned to look like a close centre system. The frame of tree prepared like this proves to be more formidable and looks good. *See* training of young trees.

Modifying genes Genes which effect the expression of a major gene but which have no effect in its absence.

Moisture test Determining the per cent of moisture in the seed by electric tests, the Brown-Duval or oven methods.

Moisture testing of seed samples For making rapid moisture determination, the equipment is available in the laboratories which do the seed testing work. The universal moisture testers are both portable, self contained moisture testing machines which use electrical conductivity as the measure of moisture percentage present. Where a certain weight of seed is required, weighing should be done accurately.

Monocotyledon Having one cotyledon. A subclass of the Angiospermae characterised by having a single cotyledon in the seed. They are usually herbaceous plants and do not show secondary growth and are in this respect considered more advanced than the Dicotyledonae.

Monoculture Cultivation that is limited to a single crop. There are obvious risks to the farmer who depends on one crop for his livelihood. For example, the monoculture of cotton in the USA at the end of the 19th century resulted in considerable hardship when the crop was ruined by the boll weevil.

Monohyrid A hybrid heterozygous at one locus and obtained from crossing homozygous parents with different alleles at a given locus; for example, Mendel's cross between tall (TT) and dwarf (tt) garden peas to give a tall monohybrid (Tt). When a monohybrid is self dominant and recessive phenotypes appear in the offspring in the ratio of 3 : 1 (the monohybrid ratio). An offspring of parent differing in one character. Compare dihybrid.

Monohybrid ratio When a cross is made keeping in consideration only one character and the pure parents for this area crossed, the F_1 shows only the dominant gene but the F_2 shows the phenotypes in 3 : 1 ratio.

Monsoon A large scale seasonal repetition of winds, pressure and rainfall in the tropics. The largest and the best developed monsoonal area in the world is South East-Asia. During the summer (April to September), the interior is intensely heated creating a low pressure area into which winds are drawn from over the surrounding oceans. This creates a South West monsoon over India and as the air over the ocean is very moist, it results in very heavy rainfall.

Monsoon forest The tall-growing extensive forests of tropical Asia that experience a marked hot dry season. Monsoon forests resemble equatorial rainforests, the main difference being that the trees are deciduous and shed their leaves after the monsoonal rains. This avoids transpiration (water loss) in the even hooter but dry season that follows. Grasses, including bamboo, are commoner in monsoon areas, but epiphytes are less characteristic. *See* also equatorial rainforest.

Moth bean (*Phaseolus aconitifolius* Jacq.) Moth bean is the most drought-resistant of the Kharif pulses and is largely grown as a dry crop either along or subsidiary to millets or cotton. It is also grown mixed with irrigated fodder jowar and bajra. On account of its mat-like spreading habit, it is highly useful against wind erosion in sandy areas. Occasionally, it is used as green manure. Unripe pods are consumed as vegetables. The grain is used as pulses in the main producing regions. Green as well as dry plants make good fodder.

Mould board ploughs These consist of a share and a mould board of steel. These are probably one of the first implements to be introduced in India from foreign countries. They differ in action from the desi plough in that a square furrow is cut and the soil inverted either completely or partially, depending on the curvature given to the mould board plough. In the case of long beam ploughs, beam and the handle is made up of wood, whereas in the case of short beam plough, the complete implement is made of metal, where it is necessary to have one way of ploughing, the turnest type of ploughs are used.

Mulch Any material applied to the surface of a soil primarily to prevent loss of water by evaporation or to keep down weeds may be designed as mulch.

Mulching Such inter-culturing operation kill weeds and produce a two or three-inch thick soil mulch which helps to reduce moisture evaporation from the top soil. Inter-culturing also breaks the surface crust which forms after rain storm. Recent investigations in this country, particularly in dry farming areas, have amply provided the protective value of both soil and vegetative or stubble mulches. The latter also increase the soils supply of organic matter and improves water absorption.

Multiple cropping The growing of a succession of crops during the year on the same plot pf land. This is only possible where the growing season is continuous or extends over most of the year and the fertility of the land is maintained. Double cropping of rice is common in S. China; in Indonesia and other parts of the tropical world, three crops are possible in some areas.

Multi-unit cooperative societies Their administration is the responsibility of the Central Government. The Act governs the working of the cooperative societies whose objects and areas of operation extent to more than one state.

Muriate An obsolete name for a chloride; *e.g.* muriate of potash (KCl).

Muriate of potash This is soluble in water and contains 50–60 per cent potash. It is a grey crystalline containing 50 to 63 per cent potash (K_2O), the whole of which is readily available. Though highly soluble in water, it is not lost from the soil as it is absorbed on the colloidal surfaces. It can be applied at sowing time or prior to sowing.

N

Napier or elephant grass *Pennisetum purpureum.* A high yielding perennial grass. It grows well in warm and moist conditions on variety of soils-not suffering water logging. The grass is propagated vegetatively from two budded sets or rooted stocks. It is a heavy yielder. Once established, this grass gives fodder for a number of years. Hybrid napier fodder is more juicy, succulent and nutritious than the common napier and its yield is high.

Narrow based ridge terrace This is commonly called a 'bund' and consists of comparatively narrow based embankment constructed at intervals across the slope of the land either on a level that is along contour, or slightly graded longitudinally, if the conditions require removal of excess water, soil, rainfall, etc. The vertical spacing between bunds may ordinarily vary from three to six feet depending upon the top type of soil, rainfall, etc. These bunds are constructed on agricultural lands upto a slope of about 10 per cent.

Nasturtium indicum Fa. Cruciferease. An annual winter season weed. Same as for *Brassica sinensis*.

Natural hazards The largely unpredictable sudden change in the environment which may have disastrous effects on human activity. These include earthquakes, volcanic eruptions, floods, draught, storms, hurricanes, and diseases of human beings, animals and plants.

Natural mulch It is formed from soil itself. It is to interrupt capillary transfer of water. (1) due to a lack of contact with the moist horizons underneath, (2) due to an increase in the noncapillary pores and a reduction in those favouring capillarity, and (3) due to the resistance to wetting that most dry soils exhibit.

Natural resources These sources of wealth, which range from soil to fissionable material, can be classified according to the nature of their supply: non-renewable or wasting resources include fossil fuels, chemicals and minerals; continuous resources include solar energy; while flow resources are those which may be depleted, sustained, or increased, and include forests and soils.

Neburon A herbicide. 1-n-butyl-3-(3,4-dichlophenyl) 1-methylurea.

Necrosis It has been seen that the mangoes planted near the brick-kilns start getting blackened at the shoot-tips. This blackening is necrosis. This damages the foliage of the tree thus reducing the yield of fruits.

Nectarine It is a small smooth skinned peach with a richer flavour. Nectarine may come out of peach seeds as also peaches from nectarine seeds or either may originate as a bud mutation from the other.

Neophyosis In citrus, it has been seen that varieties that have been propagated vegetatively for many generations become somewhat lacking in vigour. Scientists have pointed out that such varieties may have their vigour restored by being grown once from apogarnic seedlings. For that process of rejuvenation the word neophyosis was coined.

Neutral fertilizer effect The average mixed fertilizer is called neutral as it does not leave any residual effect on the soil to which they are applied to, or do not influence the soil pH. For instance, the average mixed fertilizer had an acidity equivalent a ton about 150 lbs. of calcium carbonate. This is obtained by adding dolomite limestone to the fertilizer in amounts sufficient to offset the acid residue.

Neutralizing power of lime It is used as a statement of strength in terms of calcium carbonate, that is, its $CaCO_3$ equivalent. Thus by dividing by the factor 0.56, the calcium oxide equivalent of say 102 becomes 182.1 the CaO equivalent. This means that every 100 Ibs. of the impure burned lime is equivalent in neutralising capacity to 182.1 lbs. of pure calcium carbonate. The conversion factors for lime calculation are in vogue already.

Neutralizers Such materials which reduce acidity in fertilizer mixtures such ad dolomitic limestone, are the common neutralizers.

Niger (*Guizotia abyssinica* L.f.Cass.) The crop is said to be indigenous to tropical Africa and more especially to Ethiopia. The crop is grown for seed used for oil extraction. The seed contains 38 to 50 per cent pale yellow, odourless, edible oil with a pleasant nutty taste. Oil of inferior quality is used for soap making, lighting, lubrication and as a drying oil. Seeds are sometimes used for making Chutney. Niger cake is used for feeding milch cattle.

Night soil as manure Since human excrement is a potential source of soil improvement, public health authorities in many towns make necessary arrangements for its conservation and conversion into a form in which it can be safely employed as a manure. Dehydration of night soil as such or after admixture with absorbing materials like soil, ash, charcoal, saw dust etc., produces a poudvette that can be easily used as a manure.

Nimbus Denoting clouds from which rain is falling (*e.g.* cumulo nimbus, nimbostratus).

Nitrates in rain Nitrogen occurring in rain water generally is in nitrate and ammonical forms, and consequently is readily available to plants.

Nitrate nitrogen absorption Nitrate nitrogen of the soil may go in four direction. It may (1) be synthesised by micro-organisms (2) by higher plants (3) be lost in drainage or (4) leave the nitrogen cycle in a volatile condition.

Nitrate of soda *See* inorganic ammoniates. It may if used year after year in large amounts, not only produce alkalinity but also develop an unfavourable physical condition, due to the dispersive and pudding effects of the sodium carbonate upon the colloidal complex. This has been an important inorganic source of nitrogen. It is popular because it supplies nitrogen in the form that immediately stimulates the crop even if the soil is cold.

Nitrate reduction This reduction of nitrate to the elements form was thought to be the main reason for volatilisation of this element. The trend of the reaction may be:

$$NO_3 \rightarrow NO_2 \rightarrow NO \rightarrow NH_2OH \rightarrow N$$

Nitrate-Nitrite-hypo-nitrites-hydroxylamines-Gaseous nitrogen.

Nitre *See* potassium nitrate.

Nitrification in rich organic soils The nitrate accumulation even in a low lime-organic soil usually is far greater than that of mineral soil. This is because of the ready nitrifiability the organic constituents of, the presence of adequate calcium and the inactivity of parts of carbon.

Nitrification and pH of soil Nitrogen fixation and nitrification take place vigorously in soils only having pH values about 5.5. Decay, aminization, and ammonification although curtailed, will still proceed with considerable intensity at lower pH values because most fungi are aple to effect these enzymic transfers.

Nitrification and nitrogen-carbon ratio The-addition of sugars, celluloses and other easily decomposed substances to the soil, a wide nitrogen-carbon ratio causes, if conditions for decay are favourable, a disappearance, wholly or in part of the nitrates originally present.

Nitrification and soil condition The nitrifying bacterial are not a special purpose type, but they extremely sensitive to their environs and most of the soil conditions where they live. They are much more influenced than the aminizing or ammonifying bacteria. As a result, nitrification is often spoken of as the weak point of nitrogen cycle.

Nitrification and soil temperature Nitrification does not begin before soil reaches a temperature of about 40°F, the most favourable limits being 80°–90°F.

Nitrification Being biochemically important the soil bacteria participate in all of the more common transformations, besides supporting certain special changes. Nitrogen fixation, nitrification and sulphur oxidation are essentially bacteria monopolies. These are simplest and most numerous of soil life. This is conversion of ammonia to nitrite and nitrite to nitrate carried out by certain bacteria in the soil.

Nitrification process Amino acids and similar nitrogenous materials readily yield ammonium compounds by enzymic hydrolysis. This transformation is ammonification and is brought about by a large number of heterotrophic organisms-bacteria, fungi and actinomyces. The ammonium ion in soil solution combines with carbonate or bicarbonates anions. In this form, it is readily available to small and higher plants. In favourable the ammonium ions are readily oxidised by two organisms–nitrite and the nitrate bacteria.

Nitrifiers *See* ammonification.

Nitrogen simple forms in manure In a manure heap when decay changes from aerobic to anaerobic, it becomes slower and the temperature is lowered. Carbon dioxide is still evolved in large amounts, but instead of ammonia, the nitrogenous matter is converted under ideal conditions for biochemical reduction amines, amides, ammonium, nitrite and nitrates are readily affected in this way.

Nitrogen This and other 12 elements are made available to the plants from the soil. Nitrogen has an important place among all the elements as it is used by the plants in greatest quantities. In plant tissue, it is present in 1 to 3 per cent. In plant proteins, it is about 16 to 18 per cent. This element is used by plants in the form of NO_3 or NH_4.

Nitrogen volatalization It is common in manure as well as in soil.

Nitrogen fertilizers According to the manner in which their nitrogen is combined with other elements, the nitrogenous fertilizers are divided into four groups: nitrate, ammonia and ammonium salts, chemical compounds containing nitrogen in amide form, and plant and animal by-products.

Nitrogenous fertilizers *See* inorganic-ammoniates and organic ammoniates.

Nitrogen-carbohydrate relationship Chemical studies on plant tissue as well as fertilizer experiments have shown close relationship of abundant nitrogen with vegetative growth. In the same way, the presence of large amount of carbohydrates in plant tissue accompanies fruitfulness. It has also been observed that plants deficient in either nitrogen or carbohydrates in the tissues were weak in growth and unfruitful, those with an abundance of both on with growth and unfruitful, while those with an abundance of carbohydrates and a smaller amount of nitrogen were fruitful but less vegetative.

Nitrogen-carbon ratio and climate Climatic conditions, especially temperature and rainfall, exert an important influence on the amounts of nitrogen and organic matter found in soils. Moving from warmer to cooler climate, even the rainfall may not be exactly the same, the organic matter and nitrogen of soils tend to increase. At the same time, nitrogen-carbon ratio widens.

Nitrogen-carbon ratio and nitrate accumulation *See* nitrate accumulation.

Nitrogen-carbon ratio minimum values This ratio somewhat above that of micro organic tissue, ranges for surface soils, from 1:8 to 1:18, the median being between 9 and 12. The variations seem to correlate closely with climatic conditions, especially temperature and the amount, and distribution of rainfall.

Nitrogen-carbon ratio's practical importance As long as carbon-nitrogen ratio on the soil organic matter is wide, the activity of the general purpose decay organisms will be intense and the demand for nitrogen great. Little ammonicaL nitrogen will be available for nitrification and nitrates will appear sparingly, if at all.

Nitrogen content of important fertilizers Some of the principal nitrogenous fertilizers are those that constitute the organic ammoniates.

Nitrogen cycle In all soils, there is a very considerable intake and outgoing of nitrogen all the year round. This is accompanied by many complex transformations, some of which can be controlled and created by man while others are entirely beyond his command. This interlocking succession of reactions, reversible, infinitely recurrent and largely biochemical, constitutes what is known as nitrogen cycle. The circulation of nitrogen between organisms and the environment. Atmospheric gaseous nitrogen can only be used directly by some bacteria (*e.g.* Clostridium) and certain blue green algae (*e.g.* Nostoc). They convert nitrogen to ammonia, nitrites and nitrates, which are released into the soil by excretion and decay. Another method by which atmospheric nitrogen is fixed is by lightening. Most plants can only use nitrogen in the form of nitrate, except those that are in symbiotic association with Rhizobium bacteria for other root nodule forming organisms.

Nitrogen cycle's two main divisions These are assumed on the basis of two different activities (1) synthetic or building up process (2) transformations such as ammonization, ammonification, nitrification and reduction, all resulting in progressively simple products.

Nitrogen fixing bacteria Within the last century, the benefit of improving the soil has been traced to the fixation of nitrogen through the agency of bacteria contained in nodules on the roots of the host, leguminous plants in this case.

Nitrogen forms in soil It occurs as organic-combinations, proteins, amino-acids and similar forms colloidal or rather easily decomposed.

Nitrogen in available forms Since the soil nitrogen is an essential constituent of the organic matter, the decomposition of the latter must take place in order that the former may appear in simple and available forms.

Nitrogen increase due to green manuring of soil It has been reported that an average crop of alfalfa 200–250 lbs. of nitrogen per acre are added by an average crop. Similarly 100–150 lbs. of nitrogen can be added to the soil by an average crop of clover.

Nitrogen ionic forms of soil They are NH_4^+, NO_2^-, NO_3^-.

Nitrogen in farm manure It is about 0.5 per cent in farm manure.

Nitrogen in soils In all soils, there is very considerable intake and outgoing of nitrogen in the course of a year, accompanied by many complex transformations. Some of these changes may be controlled more or less by man, while others are beyond his control.

Nitrogen loss due to farm manure Decomposition both an aerobic and anaerobic nature will cause a rapid waste of nitrogen in the manure being rather susceptible to such losses.

Nitrophos It has 20 per cent nitrogen and 48 per cent phosphorus.

NPA-N-L-naphthyl-phthalamic acid A herbicide.

Nocturnal Active at night.

Nodudlar nitrogen The nitrogen fixed by the nodule organisms may go in three directions. It may be absorbed by the host plant, the latter benefiting largely by the association. Second, the nitrogen may pass in some way into the soil itself, either by excretion or by shedding away by roots and their nodules. Third, when the legume crop is turned under, some of the nitrogen becomes available to the next crop.

Nodule organisms While the nodule bacteria are considered to be common to all leguminous plants, it is well adapted to some and not to others. The organisms from one species of legumes are not necessarily well adapted to the ready production of tubercles on other leguminous species although infection may be affected by proper culturing.

Non-essential elements of the soil While apparently sodium itself may seem at least partially non-essential, but it has sometimes to play a role to replace potassium as plant nutrient. Sodium salts may sometimes be used with some effect whenever a deficiency of potassium occurs. Marked increase in the growth of certain crops from application of sodium salts have been reported.

Non-selective contact herbicide Through coverage of all vegetation with a sufficient amount of toxic agent is needed to kill annual weeds, re-treatment is necessary in the case of perennial weeds. Some examples are ammate, diquate, DNBP General, PCP and sodium arsemite.

Non-selective herbicide for foliage These are also of two types (1) contact and (2) translocated.

Non-symbiotic nitrogen fixing bacteria Azotobacter and other non-symbiotic nitrogen fixing bacteria work independently of any host crop. Under optimum laboratory conditions, Azotobacter has been found to fix the equivalent of as much as 200 pounds of nitrogen in two million pounds of soil in three weeks.

Normal curve (Sta) It is determined by two constants representing the population *viz.* the mean and the standard deviation. In other words, knowing the mean and the standard deviation of a population, the curve can be drawn.

Normal distribution (Gaussian distribution) A symmetrical frequency curve. The mean, the median and the mode of all coincide at the same central point in the distribution. *See* frequency curve.

Normal probability integral For a chance event and far any event, the proportions in which we expect the alternative events to happen signify, the chance of the probability that they will happen.

Normal seedlings Seedlings having healthy, well farmed structures, such as epicotyl, one or two cotyledons, hypocotyl and root potentially capable of producing a healthy plant under favourable growing conditions.

Notching It is a method of pruning and is especially used in fig. It results in long stems with a cluster of branches near the end. In a short time, the trees become straggly (dispersed) and much of the fruit is borne from the trunk. It is a method used to bring about fruitfulness. In this case only one bud at a time is affected. A notch is cut through the bark and sometimes into the wood either just above or just below the bud. Notching below the bud tends to cause a concentration of carbohydrates in the bud, while above the bud is thought to increase vegetative growth by interfering with the flow past the bud of water and nitrogen.

Notification of grade specifications Grade specifications have been noticed for over 60 agricultural commodities covering over 235 trade descriptions and over 760 grading centres have been organised. Grading under Act is voluntary for commodities consumed within the country but for certain commodities, it has been made-obligatory for purposes of export.

NPK Commonly referred to nitrogen, phosphorus and potassium in agricultural terminology.

Nuclear power The power obtained from a nuclear reactor when radioactive uranium atoms disintegrate in a fission chain reaction producing enormous quantities of heat. This heat is used to raise steam (in a conventional boiler) which drives electric generators and produces electricity.

Number of class intervals This number should be appropriate to the number of items present in the series. It should not be less than 5 nor more than 20. Use of Sturgs' rule namely:

$$k - 1 + 3.3 \log 10^N$$

where, k is the number of class intervals and N is the number of items. Care should be taken, however, to avoid a bimodal distribution.

Number of knots On cotton fibre. If the knots present on the fibre are uniformly distributed, then the fibre is more strong.

Number of replications (Sta) When planning a trial, it should be kept in mind that the estimate of error variation should be based on an adequate number of degrees of freedom. If the degrees of freedom for the errors are few, the F ratio required for significance is large and it will not be possible to detect relatively small difference between treatments, in other words, the test will be insensitive.

Nurseries The initial seeding or rearing saplings of plants, especially of fruit plants, flowers and vegetables, for different uses before their final planting is done on the permanent place in the field. These are the young plants in good numbers grown at the place the nursery as small seedlings or saplings.

Nutrition losses during manure production A certain amount of every crop is lost before it is consumed by animals. Such loss while important, is usually small on the farms, especially when compared to the nutrients retained by the livestock. Therefore attention is directed towards those losses sustained by the food as it undergoes normal digestion.

Nutrient balance As the kind of fertilizer to apply and the amounts per acre are always dependent on so many things, and have to be applied a guess. It is always difficult to work out a nutrient balance. Nevertheless, the ideal should be tried for even though it seldom can be closely approached.

Nutrient deficiency in plants The deficiency of certain nutrient in a plant can be observed visually. But to be accurate in judging this one needs some experience because some other factors also show the same symptoms such as-inclement climate, factors related with the soil conditions, bad cultivation methods, bad drainage of soil or attach of disease and insect pests.

Nutrient elements as solutes These are presented to the plant both in ionic and molecular forms, the proportions depending on conditions such as in arable soil, especially of humid regions, the ionic intake is thought to predominate.

Nutrient relation to fertilizers The primary objective in the application of a commercial fertilizer to a soil is to supply plant nutrients, but the addition is usually made to supplement the already existing constituents rather than to constitute the only source of supply. Another consideration is that the fertilizer applied should establish a proper balance between the necessary elements of nutrition.

Nutrient sprays Most deficiencies may be readily controlled by applying salts of the elements to the soil or in some case by spraying them on the foliage. Injecting the elements into trees has also been tried with some success as in case of walnut mottling was controlled by infecting dry manganese sulphate into the large limbs as well as by spraying with the solution of the salt. In field crops like wheat and pulses etc. this is a common practice.

Nutrition of crops and fruit trees is a matter of supreme importance. The trees and plants remove large amounts of nutrients from the soil then agriculture can be maintained in a profitable bearing only by adequate manuring. As maintaining fertility is an expensive part of cultivation, it can repay the farmers to base their programmes on known principles of manuring and crop nutrition.

O

Oats A hardy cereal that can be grown in cooler and wetter climates than wheat or barley. It is used as an animal feed and in the form of oatmeal for porridge and other foodstuffs.

Objectionable weed seed Seed from any plant considered to be extremely destructive or harmful to agriculture. These seeds are designated by law.

Objectionable weed seed examination Sample is immediately examined for objectionable weed seed.

Oblong perforations of seed processing screens Oblong perforations are measured in the same way, except that both dimensions must be given *i.e.,* width and length with some exceptions, the width of the opening is measured in 64th of an inch. However, openings whose width is smaller than 51/2/64 are measured in fraction of an inch. The first number of the size listing of a slotted or oblong screen perforation is the width of the opening; the second number is the length of the opening, *e.g.* a 12 × 3/4 screen has oblong perforations 12/64-inch wide and 3/4 inch long.

Ocimum canum Control same as *Leucas aspera*.

Official seed samples These are those samples drawn by a seed control or seed law enforcement officer. They are submitted to the laboratory to determine if a seed lot being offered for sale meets the requirements of the Seed Act.

Oil cakes Groundnut cake meal is a useful vegetable protein food, particularly if supplemented by minerals. It is valuable for mixture with cereal and cereal by-products Mustard cake, coconut cake and linseed cake may also be used as protein supplements.

Oil cakes as fertilizers *See* cakes.

Oil engines The internal combustion engines provide both large and small units; the larger ones use kerosene, crude or diesel oil and the smaller ones petrol, kerosene or crude oil. In petrol engines, a mixture of petrol gas and air is ignited by an electric spark. In diesel engines, the mixture of oil and air is ignited by the heat created by the compression of the mixture in the cylinder of the engine. Diesel engines give dependable service, have a high efficiency and are cheap to run, but require more attention than petrol engines. The cost of petrol engines is, however, relatively less. Diesel engines require an auxiliary source of power, *e.g.* a storage battery, to start them as they are harder to start than petrol engines. They have a much higher compression ratio and are consequently made stronger.

Oil palm A species of palm tree (*Eaels guineesis*). The heavy orange-red clusters of fruit borne by the tree are the source of two vegetable oils; palm oil obtained by boiling the fleshy part of the fruit, and palm-kernel oil obtained from the soft inner part of the kernel. Both oils are used in the manufacture of margarine, soap and cosmetics.

Oilseed crops Similarly, groundnut is the leading oil-seed crop and cotton the most important fibre crop. Both of them are also the principal cash crops in many parts of the world. Amongst 'miscellaneous crops,' black pepper, chillies, tobacco, tea and potatoes occupy large areas.

Oil seeds and oil cakes Oilseeds and oil cakes are oil and protein-rich foods. They are highly palatable, easily digested and are generally used to balanced ration of farm stock in respect of protein. They are rich in phosphorus but their calcium content is low.

Okta A unit used in meteorology to measure cloud cover. It is equal to a cloud cover of one eight of the whole sky; eight oktas are therefore equivalent to a total cloud cover.

Olive The fruit of an evergreen tree that grows in Mediterranean region. The fruit is picked in the autumn and can be pickled or canned. Most of the fruit is pressed to yield a rich oil, which can be used for cooking as an alternative to butter. Olive oil is also used in soap and cosmetics.

Open centre type of tree The tree is given the vase shape. *See* vase shape and training young trees.

Open ditch drainage Two types of drainage can be taken up-open and closed-Ditches have the advantages of large capacity and are able to carry water at a low grade. Closed or under drain should be preferred.

Optimum moisture for plant growth It is evident by experiments that the most favourable moisture condition for grown of higher plants and also for moist micro-organisms as well. Occurs in soil when moisture is present in large enough amounts to be fairly low tension preferably in the neighbourhood of pH 2.7.

Optimum plant population Proper row and seed spacing of any crop leads to the growing a desirable optimum number of plants per unit rea.

Orchardist A person who works in an orchard (fruit garden) to produce fruits and vegetables.

Organic acids *See* amino acids and carbonic acids.

Organic ammoniates Organic fertilizers among which are tank-mud, various oil-cakes, compost fish scrap, bone-meal etc must undergo aminization, ammonification and nitrification before their nitrogen becomes fully available to higher plants.

Organic carbon *See* nitrogen carbon ratio.

Organic combination of phosphorus When phosphorus is held in organic combination, decay will greatly encourage its simplification. The mineral phosphorus does present a much more difficult problem. The various soil phosphates are usually rather insoluble and even when the normal solvent agent, carbon dioxide is supplied in large amounts, the rate of solution is slow. The situation is made still more difficult by the small amounts of the element present.

Organic manures These are those manures which are prepared by materials of organic resources such as green manure, dung manure, compost, blood manure and oil-cakes etc.

Organic matter Plant and animal residues, including their castings such as dung and urine etc. all constitute the organic matter. In soil formation, the presence of decomposing organic matter is a sign that the synthetic processes are well begun and augmenting those of a purely chemical nature. Under suitable climatic conditions, both plants and animals soon get a foothold on the weathering soil materials and also generation of organic residues, are left to decay and mix with the decomposing mineral mass. Organic matter changes from complex to simple products.

Organic nitrogenous fertilizers These include plant and animal by-products, such as oil cakes, fish manure and dried blood from slaughter house. Before their organic nitrogen can be used by crops, it is converted through bacterial action into readily useable ammonia nitrogen and nitrate nitrogen. These fertilizers are, therefore, relatively slow acting but they supply available nitrogen for a longer period. Furthermore, they may also supply small amounts of organic stimulants that they may contain, or of some of the minor elements needed by plants.

Orientation The lining of a map or place table survey in the field with visible objects so that magnetic north on the map or survey is correctly positioned. Orientation is necessary in plane table surveys when the table is moved to the other end of a fixed base line and when it is moved to a new station. The table is oriented by slighting back to the previous station whose position had already been determined.

Original home of banana Warm moist region of tropical Asia.

Original home of plants The place of origin of any organism is supposed to be the place where it is represented by the highest number of variability.

Ornamental gardening Gardening being an work involving certain kind of art also and thus involves a scheme of arrangement of plants, flowers in a efficient, scientific and systematic manner. It is not a simple collection of plants and flowers.

Ornamental plants Such plants which are meant to decorate and beautify the garden, house or terraces. They are beautiful in foliage, flowers and stems.

Ostiole The passage to the interior of the fig fruits, the syconium.

Otahaute gooseberry Star gooseberry, *Phylanthus acidus*. Family Euphorbiaceae.

Other crop seed in testing These include seeds of plants grown as crops. The classification of immature, damaged, diseased and empty seeds of other crops is the same as that provider for pure seed.

Overhead method of irrigation *See* aerial method of irrigation.

P

Packing Fruits should be packed in such a way that they are not hurt or bruised during transit. Size of pack has to be decided according to the fruit size and the grower's experience. Generally fruits are packed in light wooden or bamboo baskets made of Arhar stems. Ir any case, the boxes or the baskets should have sufficient lining so as to avoid damage as far as possible.

Packing fruits and vegetables In transporting, fruit by rail packing is a problem of great importance and great difficulty. The most common method now in this country is the basket. There are many forms of baskets used. These are made of several materials such as bamboo and stem of pigeon-pea and date-palm leaves etc. Often two are used, one from lid of another. Grass and leaves are frequently used to line the baskets and sometimes used between the layers of fruits. Boxes are also used for packing fruit only to a very limiter extent. Although they have very definite advantages aver baskets. A well made box is sufficiently rigid to protect the fruit, even if at the bottom of tall pile.

Paddy Unhusked rice; the term rice is not used until the husk has been removed. There are two main forms of paddy that are grown on irrigated fields and upland rice, which is grown without irrigation where the rainfall is sufficiently heavy. Paddy is the staple diet of over one quarter of the world's population.

The heaviest cropping regions are the deltas of SE Asia, *e.g.* the Irrawaddy, Jianaxi, Red River and the Yanetze (Malay).

Paddy hullers Paddy hulling or shelling has been done in India by means of a *dhankli* or a wooden lever. The Khadi and Village Industries Commission has introduced a Rajasthan type of huller which is quite efficient. Recently a manufacturer has put into market hand operated hullers similar to Japanese types. They are operated by two persons by means of levered handles. The shelling and hulling is done by the impact of the grain on a very fast revolving rubber disc.

Paddy sheller *See* paddy hullers.

Paddy threshing Threshing in the case of paddy, if the area is small, is done mostly be beating. In the case of other crops, it is done by treading the crop by means of teams of animals. Stone rollers about 3 in length and 1 to 2 in diameter have also been used in some areas. However, no other thresher has become as popular as the "Olpad thresher" which consists of 14 to 21 serrated discs fixed on a rectangular wooden frame in three lines. A seat is provided for the driver and also a guard. The implement is pulled by a pair of bullocks and worked on the harvested material, which is spread thick on the threshing floor. Threshing with this thresher is not only cheaper but also more efficient and expeditious than treading by other methods.

Paddy weeders For some special crops-such as for paddy, special types of paddy weeders have been designed and are being used and manufactured in India. They are based on the Japanese design. These consist of two rollers with pointed fingers and are pushed in between rows of paddy plants when there is standing water on the field. The idea of using these weeders is not to uproot the weeds and bring them to sufrace as in case of other weeders, but to bury the weeds under the mud so as to suffocate them and purify, bringing out nitrogen from the organic matter. This nitrogen is very useful to the paddy plant.

Palea The tiny upper bract which with the lemma enclosed the flower in grasses.

Paniala *Flacourtia cataphracta*. Fam. Bixineae (Rufiacea or Florcurtaceae).

Panicum antidotale A very promising fodder grass.

Panty price A price that makes it possible for the seller of a unit of an article to buy as much of other things and services as he used to purchase with a unit of the same article in a given base period.

Papain When the skin of an unripe but mature papaya fruit is broken, a white juice oozes out which contains a large amount of enzyme papain, which is able to digest protein. It is very similar to animal enzyme pepsin. Its importance is the manufacture of commercial papain, which is used in medicine and for other purposes.

Papaya (*Carica Papaya* I.) Fam. Caricaceae. Papaya occupies a very small area, yet its cultivation is spread throughout the country. It grows well amongst everywhere except at altitudes higher than 1,500 metres. It cannot tolerate low temperature. A dry warm climate is necessary. Strong winds are highly detrimental as the hollow stems break easily. Even though it is adopted to a wide range of soils, it grows best in a loamy soil. Deep clay soils that are prone to water logging should be avoided. In heavy rainfall areas, prolonged standing of water near its stem is highly injurious.

Papaw The name Papaya is supposed to be corruption of the Carib name for the fruit *ababai,* which has given rise to similar names in other languages. It probably accounts for the English name Papaw, but the later should not be used, as it is also used for a minor fruit of southern U.S.A., the *Asimina triloba*. The term tree-melon is also sometimes used and is not entirely unjustified as it resembled to some extent the melon to which it is distantly related.

Para grass *Brachiaria mutica* This is water loving perennial grass. It is propagated exclusively by rooted-cuttings with two or three nodes in each set. Being potentially more productive than other grasses, a heavy manuring is required to set good results in the form of grass yield. The grass is very palatable in all stages of its growth, its protein contents, are slightly lower than the other grasses.

Paragola system of grape vine training In the paragola, arbour or bower system vine is allowed to develop into two or three branches about three feet from the ground. The branches are fastened to the horizontal wires of the paragola and allowed to

Paragola or Gate System of Training a Grape Vine

grow on the arms, are pruned each year according to the mode of bearing of the variety planted.

Parallel drainage A drainage pattern in which the mainstreams are tributary streams follow virtually parallel courses. This develops where there is a strong structural control in one direction; for example, where strata are gently dipping.

Parameters A collective name given to statistical measurement as a mean, median, modem, mean deviation, stand and deviation.

Parshall flume and V-notch Control of water in the field is also important. Water flow can be easily checked by canvas dams in small irrigation channel. Spills made of bamboo, wood or tiles and diversion boxes efficiently control water flow from fed channels to furrows or borders. These simple and inexpensive devices are labour saving and increasing irrigation efficiency.

Parkinsonia aculeata An important hedge plant and is fast growing.

Pasting of nutritive elements This is another method of detect elemental deficiency in plants. The rules for application are the same as for sprays. Pasting is done only on half leaf so that the difference may be noted then and there for the same leaf.

Pasteurization The partial sterilization of food stuffs by heating to a temperature below boiling. This kills harmful micro-organisms but retains the flavour. It is named after the pioneer of the method Lewis Pasteur, who used it to prevent spoilage of wine and beer.

Patch budding It is similar to shield budding but in it the section of bark containing the bud is rectangular and the bark of the stock from an area the same size and shape, the bud is carefully befitted in this place.

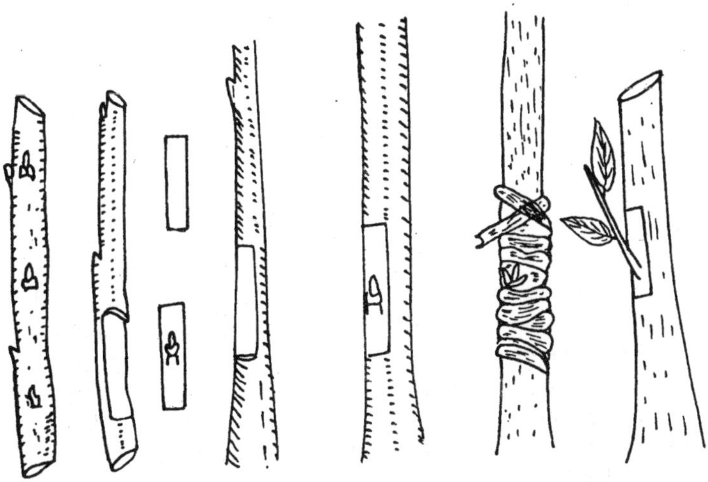

Patch Budding

PCI *See* per capita income.

PCP Pentachloro, phenoll. A herbicide.

Peach (*Prunus persica* Batsch) Fam. Rosaceae. The area under peach is very small and is mainly located in the Himalayan hills at various elevations.

Peach bearing habit Fruit buds are borne laterally upon one year old wood and on short spur like twigs. Ordinarily, they develop two fruit buds and a leaf bud at one node. The fruit buds are usually located from the middle of the shoot upwards. In cutting away branches, the position of fruit buds should be taken into consideration.

Peach-pruning In the case of bearing trees, annual pruning is necessary to maintain the centre open. Two to three year old branches may be cut back to the outward pointing side branches to encourage a spreading growth. Shorten and thin outside branches to stimulate the growth of new fruiting wood every year. A satisfactory annual growth should be as to 50 cm long.

Peanut *See* groundnut.

Pear guava *Psidium pyreforum*, Fam. Myrataceae.

Pear (*Pyrus communis* L.) Pear is grown mainly in hill regions at elevations ranging from 1,500 to 2,500 metres. Its cultivation is rather restricted mainly because the fruit does not store well. Most of the pear varieties are self-unfruitful and the planting of polleniser varieties is advocated. *Nashpati* is a self-fruited variety.

Pears' propagation and pruning Pears on pear stock make a vigorous growth and develop into large trees. They remain dwarf on quince stock, when they are trained into different forms. For pyramids form, cut the graft while planting at about three-fourth metre above ground. Next winter, prune the leader to about 25 cm and the laterals to about 20 cm to the outward-pointing buds. In the second summer, all the branch leaders and laterals should be pruned to five or six leaves from the clusters, allowing the central leader to grow unchecked. In the third winter, the central leader is cut back to about one-third its lengths, but the branch leaders and laterals are not pruned. In the third summer, the branch leaders laterals, except the central leader, are again cut back to five or six leaves as in the previous summer. In the fourth winter, the central leader is again cut back to one-third its length. By following this procedure, a pear tree on quince stock would start flowering in the fourth year. Rearing pear trees are pruned as in the case of apples.

Pearl millet *Pennisetum americanum*. Used as a staple grain crop.

Pecan *Carya pecan* Fam. Junglandaceae.

Pectin It is one of the essential ingredients of fruit jelly. In some fruits, the failure of the juice to form jelly is due to the lack of the substance known as pectin. This is present in most fruits but in some it is absent. It is extracted from certain fruits such as apples and citrus and may be brought as powder or solution. In the cell walls of green fruits, there is an insoluble substance called pectose, which as the fruit ripens turns into a soluble colloidal substance pectin. As the fruit ripens fully, pectin is converted into pectic-acid and methyl alcohol.

Pellet application Small balls are made of 12 parts of soil plus 1 gram of manure. These small balls are put in the soil at 1 or 2 inches depth. Such method is applied in paddy crop.

Pennisetum typhoides Commonly known as Bajra crop. It is severely attacked by the green ear disease and grain smut.

Pennisetum pedicellatum Called Dinanath grass, a good fodder grass.

Pennisetum polystychum The thin napier grass.

Pepper (*Piper nigrum* L) Pepper is an evergreen plant, indigenous to the rain forests of the south-west coast of India. Black pepper is the dry unripe fruit, while the ripe fruit deprived of its tender skin and pulp gives the white pepper of commerce. Pepper is used a spice or condiment for flavouring food stuffs. It is also used as a carminative. The alkaloid piper line forms 5 to 8 per cent by weight of seed and the volatile pepper oil, 1 to 3 per cent of the unripe berries. Pepper is highly valuable product for export.

Pepsin An enzyme that catalyses the partial hydrolysis of proteins to polypeptides. It is secreted by the gastric glands in an inactive form, pepsinogen and is activated by hydrogen ions. At pH values of 4.6 and less pepsin initiates the digestion of proteins, splitting them into smaller fragments. The extent of this action is proportional to the length of time the protein is in contact with the enzyme.

Percolation If the large pores more or less vertical and continuous and where they are numerous enough to ensure a large carrying capacity, the percolation of water through soils would be very rapid. *See* also gravity flow.

Percolation losses of nutrients *See* drainage.

Percolation losses of water When at any time the amount of rain fall entering a soil becomes greater than its water-holding capacity, losses by percolation tend to occur. These losses will be influenced not only by the amount of rainfall and its distribution, but also by evaporation, the character of soil and the presence of crop.

Percolation mechanism This may be referred to as the tension gradient. If it is high *i.e.* if the difference in tension between two points a unit distance apart is great, water will adjust toward the higher tension with comparative rapidity.

Perfect market Same as perfect competition.

Perenox A chemical weed killer.

Perennation A vegetable means of surviving unfavourable seasons, seen in biennial and perennial plants. The metabolic activities are reduced to a minimum, usually by die-back or aerial parts and food for the next growing season is stored in swollen under ground organs. Seeds may also be regarded as perennating organs.

Perennial irrigation Irrigation methods that provide water for cultivation throughout the year such as projects based on dams and underground resources that are not subject to seasonal fluctuations.

Pericarp The mature ovary wall.

Permanent preservation of fruits Sterilization, use of antiseptics, use of chemical as preservatives, fermentation and exhaustion of the containers are some of the methods involved.

Permanent buffer strip cropping In the case of permanent or temporary buffer strip cropping, the strips are established to take care of critical *i.e.* steep or highly eroded slopes in fields under contours trip cropping. These strips do not form part of the rotation practised in normal strip cropping and they are generally planted with perennial legumes, grasses or shrubs on a permanent or temporary basis.

Permissible germination seed defects Seedling of Pisum, Vicia, Phaselous, Lupins, Vigna, Glycine, Arachis, Gossypium, Zea and all species of Crucifereae with a damaged primary root but with several secondary roots of sufficient length and vigour to support the seedling in soil. Seedlings with superficial damage or decay to the hypocotyl, epicotyl or cotyledons which is limited in areas and does not affect the conducting tissue. Seedlings with dycotyledons with only one cotyledons. Seedling of tree species having epigeal germination when the radical is four times the length of the seed, provided all structures which have developed look normal. Seedlings which are seriously decayed by fungi or bacteria, but only when it is clearly evident that the parent seed is not the source of infection and it can be determined that all the essential structures were present.

Persian wheel The water is lifted in a series of small buckets, each holding about one gallon. The buckets are fixed on an endless belt or sprocket chain, which moves on a vertical wheel or drum. The bullocks walking in a circle turn the drum through a set of bevel gears and shaft, which in turn moves the endless belt. The revolving movement of the belt raises the buckets full of water on one side, which are emptied as they come to the top and lowers them on the other to fill again with water. The discharge is about 2,500 gallons per hour for a lift of 25 to 30 feet. In some models *e.g.* Noria, and engine is used to run the Persian wheel. This admits of large buckets and greater depths.

Persimmon (*Diospyros kaki* L.) Fam. Ebnaceae. Persimmon is grown at elevations ranging from 900 to 1500 metres. The promising varieties are Fuy, Hachiya and Hyakume. Several good varieties like Dai Maru and Tenanashi are also being grown successfully at the pomological stations, Coonoor on the Nilgris in south India. The tree is propagated by grafting (whip and tongue method) on seedlings of *Diospyros lotus* and *D. virguniana*. The grafts are planted in winter, 6½ to 7½ metres apart.

Persimmon planting and pruning The trees are headed back onto two metres above ground at planting time. Four to five shoots are allowed to grow all round the stem to avoid narrow crotches and to develop a well-balanced head. There is no further pruning after this. Dead, broken and interfering branches are removed every year.

Pest management strategies

1. Destroy termite mounds and treat the spots with Lindane or chlorophysiphos spray.
2. The only well rotten FYM.
3. For direct prevention of termites, treat seeds with indosulin or chlorophyriphose using 240 and 90 ml. a.i./ha over 5 litres of water and spray over 100 kg seed, spread over

floor and periodically turn over. This has to be done 1-2 days before sowing.

4. Avoid late sowing of crops.
5. If prevention treatment of any seed has been taken and termite incidence is evident in standing crop, broadcast the insecticide to treat the soil.
6. For late sown crops, seed rate should be increased.
7. For sucking insects like mites aphids and jassids spray FOFM or phosphomidon @ 160 and 250 ml a.i./ha may be sprayed.
8. To control the pest like army worms, cutworms, semiloopers, which are sporadic in occurrence, spray DDVP or fenitothion @ 500 ml. a.e./ha when one or more caterpillars are noticed. This treatment is good for *Pyrilla* control also.
9. To prevent damages, zinc phosphate baits may be applied during stage of earring in wheat and fruiting.
10. At harvest, the stubbles or crops left over should be removed and burnt to do away all after effects of insect pests and diseases.

Phalaris canariensis A good fodder annual plant is equal to oats.

Phalsa *Grewia asiatica* Fam. Tiliaceae.

Phaseolus vulgaris This genus is attacked by the bean rust *Uromyces appendiculatus*, which is widely distributed. It also attacks several other species of genus Phaseolus such as *P. radiatus* and *P. mungo.* It is a leguminous vegetable crop of which pods are used.

Phosphorus Phosphorus influences the vigour of plants and improves quality in crops. It encourages formation of new cells, promotes root growth (particularly the development of fibrous roots) and hastens leaf development, emergence of ears, formation of seed and grain, and maturation of crops.

Phosphoric acid availability Those fertilizers which have water soluble or citrate-soluble compounds of phosphorus have this element present in available forms such as (water soluble) $(CaH_4,PO_4)_2$. $NH_2H_2PO_4$ and K-phosphate and citrate soluble IN 15 per cent solution of citric acid $(Ca_2H_2(PO_2)_2$.

Phosphoric acid unavailable forms These are phosphates of bone and raw phosphate rock. Phosphoric fertilizers are:

Fertilizers	*Per cent Elements available*
Raw bone meal $Ca_2 (PO_4)_2$	22–25
Steamed bone meal	23–30
Rock phosphate four and chlor-apatites	25–30
Superphosphate $CaH_4 (PO_4)_2$ and $Ca_2H_2 (PO_4)_2$	16–47
Bone black $Ca_3 (PO_4)_2$	32–35
Monocalcium phosphate $CaH_4 (PO_4)_2$	50
Calcium metaphosphate $Ca_4(PO_4)_2$	62–63
Ammoniated superphosphate $NH_4 H_2SO_4$, $Ca_2H_2 (PO_4)_2$ $Ca_3(PO_4)_2$, $(NH_4)_2 SO_4$	18 (3–4 per cent N)
Ammo-phos $NH_4H_2PO_4$	48–53 (11 per cent N)
Potassium phosphate K_2O PO_4	32–53 (30–50 per cent)

Phosphorus as a fertilizer problem This element in soils is not only low but rather unavailable. As a consequence, even when soil conditions are favourable, especially as to pH and active lime, this element is likely to be critical.

Phosphorus and calcium Deficiency of calcium and phosphorus are matters of vital importance in cattle nutrition and for the proper assimilation and use of these materials, the animals must have sufficient vitamin D which is normally obtained from sunlight either by exposure to it or through sun-cured feeds, such as hays.

Phosphorus deficiency or aphospborosis Iloccurs in cattle fed largely on-rough-ages, such as hay, straw, pasture, com silage, beet pulp and molasses. This is particularly so when the roughage is grown on phosphorous deficient areas or the grains consumed by the animals are not adequate in quantity or the deficiency may be due to both the factors.

Phosphorus forms in soils Phosphates of Ca, K etc. and soluble organic forms, ionic forms as $PO_4^-HPO_4$ and $HZPO^-_4$.

Phosphorus in farm manure It is present in the farm manure from trace to 0.80 per cent.

Phosphorus influence of pH If the pH of a mineral soil is raised much above 7, say 7.5 or 8.0, not only is the presence in the solution of the PO_4 ion encouraged but also the phosphate nutrition of higher plants and also of microorganisms as well, is disturbed in another manner.

Phosphorus influence on plants Neither cell division nor the formation of fat and albumen take place adequately without it. Starch may be produced when it is lacking but does not readily change to sugar. As seeds do not form without its presence, it undoubtedly is concerned in the production of nucleoproteins. Its close relationship to cell division may account for its presence in grains and other seeds in comparatively large quantities. Flowering and fruiting depend entirely on it.

Phosphorus loss by leaching *See* nutrient losses by leaching.

Phyllanthus niruri Same as *E. prostrata.*

Physical condition of fertilizers If the fertilizers used in mixed ones, nitrate of soda, sulphate of ammonia and potassium chloride are most likely, because of their deliquescence, to develop an unsatisfactory physical condition. The condition of a mixed fertilizer is maintained by the use of dryers.

Physical purity The purity of a seed sample based upon the presence or absence of weed seeds, other crop seeds or inert matter. It does not include genetic purity.

Pickles Pickles of the western kind are made principally by adding vinegar with or without sugar and spices to fruits and vegetables. In some cases, the raw product is allowed to ferment and thus produce its own acetic acid. Indian pickles are of great variety, the preservative being salt, sugar, vinegar or mustard oil. Frequently two or more of these substances are combined.

Picota It is a contrivance in which a wooden beam is mounted, in a sea-saw fashion, on a vertical pillar fixed near a well. A bucket with a long rope is tied to one of the beam and the other end is counter balanced with a heavy stone. It is suitable for lifts upto about 15 feet. Toe bucket carries two to six gallons of water and the quantity of water lifter varies from 1,500 to 2,000 gallons per hour, on a lift of 15 feet with three-men working the picots.

Pico-Symbol: p A prefix denoting 10^{-12}. For example 1 picofarad (pF)=10^{-12}, farad (F).

Pictogram (pictograph) A graph based on picture symbols rather than on bars on lines to represent data. For example, the use of sheep symbols to represent a certain number of sheep or symbols of coffins to represent mortality rates.

Picturegram or pictorgram Such a diagram is attempted to show in pictorial form the relative amounts of wheat obtained from different sources and consumed in the U.K. in a certain year. The actual amounts are in the ratio 7 : 2 : 1.

Pigeon pea (*Cajanus cajan* I. Millsp.) Pigeon pea or red gram is extensively used as dal and its green pods as vegetable. It is almost in daily use to provide protein food. The green leaves and tops of plants are fed to animals or utilized as green manure. Husk of pods or seeds with part of the kernel constitutes valuable cattle feed. Dry stalks obtained after threshing are used for basket making or as fuel or thatching material. Being deep rooted, it is also planted as a soil renovator to break up the sub-soil and as contour hedge to check erosion. It is also often grown as a cover crop in plantations.

Pine apple *Ananas comosus* and *Ananas sativa* Shutt. Fam. Bromeliaceae. It is a humid tropical plant and grows equally well, both on the plains as well as at elevations not exceeding 900 metres. It does not tolerate very high temperature or frost. It grows in almost any type of soil provided it is free-draining.

Pineapple propagation Pineapple is commonly propagated from suckers or slips. Suckers which arise from underground parts of the plant are commonly used. Slips arise from the fruiting stem and crown on top of the fruits. After the fruit is harvested, stalks are cut into discs and used for propagation. Plants grown from suckers produce fruits in about 18 months, while those from slips and suckers propagated from disc cuttings take over two years. Suckers at fist cured by stripping the lower leaves and drying them in sun or under shade for about three to four days before planting. They are planted either in flatbeds where there is no danger of water stagnation or in shallow trenches which are gradually filled as the suckers grow and develop. Care should be taken to see that the bud or 'heart' of the suckers does not get buried, spacing varies from half to two metres between plants and one and a half to one and three-fourth metres between rows. Rainy season is the time for planting.

Pistacio *Pistacia vera*. Fam. Anacardiaceae.

Piston pump It consists of an air-tight barrel in which a piston moves up and down. It is useful for small flows and high rifts. It is not so extensively used for irrigation as the other two.

Pitch of screen In some seed processing machines, it has to be adjusted to enable the operator to move seed over the screen rapidly or slowly. The scalper scre n would be set at a steep pitch to hurry trash and weed seed out, while the grading screens would be set at a flat pitch to hold the seed on the screen longer.

Pit curing-tobacco Leaves are dried and cured in pits. Before putting the leaves in pits, they are sun-dried for 4-5 days.

Pit silo This is constructed by digging a circular or rectangular pit of desirable dimensions in the ground. A circular pit is preferable as packing of material for exclusion of air, which is very essential to get silage of high quality, can be done more efficiently in it.

P_1 The parent of an F_1; P_2 the grandparents. Also used to designate different parents used in making a hybrid.

Placement method According to this method, the fertilizers are placed in the soil at a certain depth but it is not necessary that it be so placed just near to the seed or plant.

Plane table An instrument used in surveying that consists of a horizontal board fixed to a tripod. Paper is fixed to the board and after marking magnetic north signs are taken to distant objects using an alidade. Direction lines (rays) are then drawn on the paper. The plane table is then moved to the other end of a measured base line where the operation is repeated, the point at which rays intersect marks the position of an object. *See* alidade.

Plantain Banana, edible only when cooked, *Musa paradisiaca*.

Plants as tools for logical control of weeds The outstanding example of biological control of plants is that of prickly pear (*Opuntia* spp.) in Australia. Other noteworthy examples are the destruction of Lantana in Hawaii and St. Johnswort (*Hypericum* *perforatum*) in America. Attempts are underway for biological control for nut-grass in Hawaii, the common blackberry (*Rubus fruticosus*) in New Zeland, *Xanthium strumarium* (a poisonous weed) in America, and puncture vine (*Tribulus terrestris*) in south west America. A somewhat different examples of biological control is the destruction of aquatic vegetation in ditches by herbivorous fish.

Plants for green manure In localities near forests, the paddy crop is often manured with green forest leaves. These are incorporating in the soil at the time of puddling. In recent years, extensive efforts have been made to plant *Glyricidia maculata* and *Sesbania speciosa* on the borders of paddy fields or other vacant spaces to provide green leaf for manuring the paddy crop. Grown from seedling or rooted stumps, planted six feet apart, each Glyricidia plant is said to give annually two cuttings of 15 to 30 lb. each of green leaf. It does well in both red and black soils. Similarly, *Sesbania speciosa* seedlings planted four inches apart from paddy borders produce 2,000 to 5,000 lb. of green leaf for manuring one acre of paddy.

Plant nutrients See essential elements. Plant roots are a part of colloidal system. The presence of higher plants rootless penetrate the soil, forcing their way into the pore and through the gelatinous fillings, the contact between the root hairs and soil colloidal matter becomes very close.

Planting Putting plants, trees in their permanent positions for purpose of cultivation and production.

Planting board A board used to plant saplings. The marks are at the notches on the board.

Planting distance The distance to be allowed between trees as to allow their maximum development, both of roots and shoots. It is seen that maximum yield per acre can be had by placing the trees somewhat closer, in most cases, than the distance at which the maximum yield per tree is found.

Planting holes The hole in which the plants are put in. It should be large enough to allow the roots to extend freely naturally or at least not to be twisted around each other. There should be ample room to work so that soil may be filled in around the roots leaving no air pockets.

Planting of banana Common method is by planting suckers.

Planting on slopes Slopes are better suited for such crops. Trees planted as windbreak do not allow the cooling of vegetation on their lee-side. Similarly, a canopy of shade trees protects the crop during periods of mild frost. Covering the ground by mats, paper, cloth or cellophane protects valuable crops from the effect of frost.

Planting plans In planting an orchard, a well considered plan should be followed. This should provide for the maximum number of trees per acre consistent with sufficient space for the proper development of each tree and convenience in orchard operations. If more than one kind of fruit is to be planted, each kind should ordinarily be in a block itself. It is obvious that large trees cannot be mixed with small ones in an orderly fashion without waste space. Fruit ripening at the same season should be grouped together for convenience in protection and harvesting. If there is variation in the soil, the blocks should be so planned as to suit the different species. *See* also planting system.

Planting season it is different for different crops especially it varies with the temperate (deciduous) fruit trees which are planted in winters and tropical and sub-tropical fruit trees and planned in rainy season.

Plant nutrients The plants required inter alia the following essential nutrients for their normal development:

Carbon	Nitrogen	Calcium
Hydrogen	Phosphorus	Magnesium
Oxygen	Potassium	Sulphur
Manganese	Boron	–
Copper	Molybdenum	–

Plant's removal of nutrients Plants differ markedly in their influence. Grass especially reducing the drainage loss of nutrients.

Plant protection These measures are important as their timely use for protection of crops against disease and insect pests naturally increases yields of crops.

Plant roots as soil organisms The life of the soil might, not ineptly, be divided into two opposite and in some respects, opposing groups (1) those that supply organic residues and (2) those that are engaged primarily in tearing them down. Plant roots are, of course, the all important representative of the first group while millipedes, springtales earthworms, bacteria, moulds and actinomyces are principals in the second type.

Plant roots and soil reaction If the contact of the root hairs and microorganisms with the colloidal surface of the soil complex is close, and apparently this may be the case, there may be a definite overlapping or integration of overlapping. Under such a condition, the H-ions generated at the surfaces of the root hairs and the micro-organisms, might effect an interchange with the nutrient cations without either of the ions emerging from the interfacial zones.

Plant roots and the rhizosphere The concentration of the microbial activity within the rhizosphere is ample. Of the organic matter being continually sloughed off by plant roots, thus providing food and energy for micro-organisms. Such biochemical phenomena cannot fail to increase greatly the rate and ease of transfer of nutrients from soil to plant.

Plant roots influence on nutrient availability The availability of nutrients is not always so easy and copious as the explanation generally given is. The proportion of the total exchange capacity occupied by a given nutrient is a very important factor.

Plant roots' contact relation with soil The rootless and the root hairs are in most instances in a very intimate contact with the soil colloidal system. The relationship is so close that nutrients are thought to pass from the soil solids to the plant without leaving the sphere of colloidal influence.

Plant root exudates The contact effect apparently is much accentuated by root exudates and by microbial activity in the immediate neighbourhood of the absorbing surfaces. Plant roots not only give off large amounts of CO_2 but also other acids as well.

Plants suitable for green manure *See* Crops suitable for green manure.

Plant's supply of water by capillarity *See* capillary water in soil.

Plant's supply of water and root extension The rate of root extension is very fast. The rate of elongation is great enough to take care of practically all of the water needs of plant growing in a soil at optimum moisture thus making the plant under this condition more or less independent of capillary adjustment for its immediate water supply.

Plant tissue More than 75 per cent of average green plant tissue is water. The dry substance is made of carbon, oxygen, hydrogen, nitrogen and mineral water. The other elements even though they may be present in minute amounts, playa vital role in plant nutrition and are important additions to the soil when the organic residues become a part thereof.

Plant tissue decomposition The chemical nature of the plant tissue is very complicated. Hence, its addition to the soil does set up reactions of a decidedly complex nature. A considerable portion of these residues in a colloidal state makes it difficult as to find out the successive changes to which they are subject to in the soil.

Plants tolerance of alkali The tolerance of crops to alkali depends on (1) the physiological constitution of the plant (2) the rooting habit. Much depends on the character of the alkali solution, the nature of the cell wall and the constitution and activity of the cell contents.

Pneumatic separator for seed cleaning *See* Aspirator for seed cleaning.

Pecan *Carya pecan* Fam. Juglandaceae.

Ploughing of orchards in large orchards, where tractors are used for power, ploughing may be replaced by the use of heavy disc harrow. This is not practicable where oxen are used. The soil may be ploughed at other times of the year if manure is applied or if the seeds become so large or numerous that they cannot otherwise be conveniently killed. Ordinarily however, it will be more economical to kill the weeds before they become large, by means of a harrow or cultivator. In the same way, the soil may be kept soft. During the dry season, it may be desirable that the orchard be cultivated after each irrigation, but often it will be more economical to cultivate less frequently. Ploughing of light shallow and medium soils should be omitted; they may instead be hoed twice in June–July or July–August, depending on the receipt of rains. Deep soils should be ploughed once in three or four years immediately after the harvest of a rabi crop. Operations should be done on contours. Patches instead with perennial grasses like kunda (*Cyperus rotundus*) and hariali (*Cynodon dactylon*) should be cleared by deep ploughing or hand digging is also a good method for dry farming.

Plough-sole placement By this method fertilizer is placed in the plough furrows at the time of ploughing. It is advantageous especially when there is shortage of moisture on the upper surface of soil.

Pluchea lanceolata A perennial weed, which controls with 2, 4-D plus 2,4,5-T mixture (brush killer) 2 lbs. a.e./100 gallons of water at bloom stage. Rest is same as *Carthamus oxyacantha*.

Plucking and processing Plucking and processing of tea is usually restricted to the terminal bud and two expanded leaves (two leaves and a bud). This is called fine or light plucking. Coarse plucking includes extra leaf or leaves.

For manufacturing black tea, the plucked leaves are kept for period of 18 to 24 hours in ventilated indoor racks, rolled for half-an-hour mechanically to break up cells, fermented or oxidised for about three hours at 80°F to develop flavour and colour, dried again in a current of air at 180 to 220°F and passed through sieves of different meshes, thus sorting out 'leaf', 'broken' and 'dust' tea. The 'leaf' tea and 'broken' tea are graded again into three grades each, namely (a) Orange, Pekoe and Flowery Pekoer and (b) Broken Orange Pekoe and Fannings respectively.

Plum (*Prunus* sp. L) Plum is grown mainly on the Himalayas hills where the following varieties have been successfully grown: Grand Duke, Early Transparent Gage, Victoria, Santa Rosa, Wickson, Beauty and Kelsey. In the Southr on Nilgiris and Kodaikanal in Chennai State, several choice varieties of the Japanese Plum (*P. salicina*) are grown. The more important of these are Rubio, Alu Bokhara, Gaviota, Shiro, Combination and Hale. All varieties, except Beauty, Santa Rosa, Gaviota, Rubio, Alu Bokhara and Hale which are self fruitful, require cross pollination from other varieties. Plums are usually propagated by shield budding on wild apricot or common peach stock. Planting, spacing, cultivation and fertilization are the same as for peach.

Polyembryony In mango, one seed may contain several embryos. Only one of which is fertilised. The others develop asexual and are thus genetically exactly like the tree of which they were borne. As these fertilized ones are more vigorous than the unfertilised embryos, the latter may never appear above ground. If several shoots appear, there is no way of being sure of which is from a fertilised embryo, but if all of the most vigorous are discarded, the chances are that in great majority of case the seedlings preserved will be asexual.

Pomegranate (*Punica granatum* L.) Fam. Punicaceae. The largest area under pomegranate, about 500 hectares is in Maharashtra, but there are small plantings in almost all parts. Where winters are cold, the tree is deciduous, but on the plains it is evergreen. A hot, dry summer aids in the production of the best fruit. The presence of many seeds and of tannin in the rind and membranes, detracts from its attractiveness.

Pomology Science of growing fruit plants/orchards.

Post emergence application Some chemicals especially are applied at the stage when the weed plants have come out of soil.

Post harvest treatment of fruits and vegetables This is done from controlling diseases and insect-pests in the storage, and damages done during transport.

Potassic fertilizers Most of the Indian soils contain a sufficient amount of potash. Potassic fertilizers should, therefore, be applied only to such soils as are definitely known to be deficient in potash or to respond to their application, such as sandy soils. They can also be applied to certain crops like tobacco, potato, onion, tomato, fruit trees, etc. to improve the quality and appearance of their produces.

Potassic fertilizers in common use These are (1) muriate of potash (potassium chloride) and (2) sulphate of potash (Potassium sulphate). These salts are important constituents of the waters of ocean and inland seas and of saline deposits derived there from.

Potassic forms in soil Original minerals as feldspars and mica have potassium salts as sulphates, carbonates, etc.

Potassic manures Such manures which have potash as their main constituent. Such as Murite of potash. Such potassic manures materials give potash as the major constituent of theirs.

Potassium It enhances the plant's ability to resist diseases, insect attack, cold and their adverse conditions. It plays an essential part in the formation of starch, production and translocation of sugars and is thus of special value to carbohydrate rich crops like sugarcane, potato and sugar beet. Increased production of starch and sugar in legumes fertilized with potash benefits symbiotic bacteria and thus enhances fixation of nitrogen. It also improves the quality of tobacco, citrus etc. With an adequate supply of potash, cereals produce plum grain and strong straw. But an excess of the element tends to delay maturation of crops though not to the same extent as nitrogen.

Potassium deficiency Plants can take up and store potassium in much larger quantities that what is needed for optimum growth and this excess uptake is known as luxury consumption. With maturity or death of plants, potassium is washed out from the plant body readily. Vegetables and legumes are particularly heavy consumers of potassium. Deficiency of potassium produces the characteristic ringing of alfalfa leaves with rows of small white spots; reddish brown discolouration of cotton leaves; drying, scorching and curling-under of leaf margins of leaves; and intravenal chlorosis and 'firing' along the edge of maize leaves. The older leaves are first affected.

Potassium ammonium nitrate It has 16 per cent nitrogen and 28 per cent potash.

Potassium ammonium phosphate It contains 12 per cent nitrogen, 24 per cent phosphorus and 12 per cent potash.

Potassium amount in soil it is found in about 2.00 per cent in humid soils and about 2.40 per cent in arid soils.

Potassium essential element An element used relatively in large amounts through soils.

Potassium fertilization problem Most mineral soils except those of a sandy nature are high in total potassium. Yet the amount of potash held in an easily exchangeable condition at the colloidal milifaces is small.

Potassium in farm manure The quantity present is about 0.40 per cent.

Potassium influence on plants By increasing resistances to certain diseases and encouraging the root system, while delaying maturity, it works against undue ripening influence of phosphorus acid. It exerts a balanced effect of both nitrogen and phosphorus. The leaves of crops suffering from a potassium deficiency, appear dry and scorched at the edges, while the surfaces are irregularly chlorotic. Adequate potassium encourages the assimilation of carbon dioxide, the utilization of nitrate nitrogen in protein synthesis, the intake of water and adequate root development.

Potassium losses due to leaching Much of this element is likely to be lost by leaching, sometimes to the extent of 50–70 lbs. per acre.

Potassium losses from farm manure It is probable that one-half of nitrogen, potash and organic matter and the least one-third of phosphorus of farm manure are lost even under fairly good methods of storage.

Potassium sulphate It contains 45–52 per cent potash. It should be used before sowing of seed in the field.

Potassium transfer to available form *See* solubility of potash by carbonic acid.

Poudrete *See* night soil as manure.

Power It is of a body's rate of doing work. Power = work done/time taken = W/t = F XS/t = F x V, (therefore, S/t = vit uniform velocity).

Power operated wheat thresher Recently there has been introduced a power-operated wheat thresher. It is worked by a tractor or an oil engine of 20 to 25 B.H.P. This thresher is being simplified so that it can also be used by an oil engine of 10 to 15 B.H.P.

Power tiller It is smaller tractor and is useful for hilly tracts.

Power units The power used for driving irrigation pumps may be furnished by (*i*) wind, (*ii*) steam, (*iii*) oil, (*iv*) electricity, or (*v*) compressed air.

Practical units of energy and work (*i*) watt-hour = 3600 joules; (*ii*) kilowatt-hour (kwh) = 1000 watt hours.

Practical units of power These are Horse Power (H.P.) = 550 lbs/ sec. Kilowatt = 1000 watts = 1.34 H.P.

Practical utilization of green manure F or green manuring of fields, good drainage should be assured. Aeration is of vital importance in the proper decay of plant residues. In places where rainfall is scanty, great precaution should be taken in handling of this crop. It. is generally best to turn under the green manure crop when their succulence is near the maximum and yet at a time when abundant tops have been produced. The ploughing should be done at a time when a supply of rain is maximum as to ensure a rapid and effective decomposition.

Prechill To place seeds on a moist substratum and hold at a low temperature, usually 10°C, for a definite period of time before transferring them to a higher germination temperature.

Precipitation efficiency index An index devised as the basis for classification of climate. It links mean monthly rainfall with mean monthly temperature to show how efficient the rainfall is in a particular temperature region.

Precocity Bring about earliness in fruit maturity.

Pre-cooling Sometimes mangoes are pre-cooled at 12–13°C with 500 ppm of Baustan to increase shelf life of the fruit.

Pre-emergence application In certain cases as-in treating soil for weed control herbicides, the chemical has to be applied in soil so that the weed seed may be killed before they come out of the soil.

Preparing seed for processing Excessive trash in the field-run seed interferes with proper processing and slows down capacity. A scalper can be used to rough-clean seed when trash content is high. If seeds have hulls, hard-seed coats, beards, excessive glumes, or occur in clusters, a huller-scarifier or a debearder may be used. A sheller must be used for ear maize, a dryer is essential when seeds have high moisture content.

Preservation It is the process of killing of the organisms present and the exclusion of all other or the provision of conditions under which the organisms cannot develop. Economically this process takes part of the crop out of the fresh fruit marked and thus helps to prevent gluts and low prices. It also supplies the fruit throughout the year, while the preserved fruit differs from the fresh, it is frequently a very palatable product and has many of the dietary values of fresh fruits. Much the same reasons exist for preservation of vegetables.

Preserves A fruit or vegetable in a heavy syrup is called a preserve. The most popular preserves are of amla, apple, bael, carrot, mango, myrobalan and petha (Ash gourd). Amla is pricked and boiled; apples are pealed, boiled and pricked; bael is peeled, pricked and boiled; carrots are scraped, boiled and pricked; raw mango slices are boiled and pricked; myrobalan is pricked and boiled; petha is cut into slices, peeled, pricked and steeped in lime water for 10 to 12 hours and then boiled. The boiled fruit and sugar (half the weight of prepared fruit) are placed in alternate layers in a vessel and allowed to stand for 24 hours. During this period, the fruit gives out excess of water and the sugar goes in solution. Usually the syrup will be of about 35 Brix. After 24 hours, more sugar is added to raise the strength of the syrup to about 60 Brix. A small amount of citric acid or tartaric acid (1 to 2 oz. per 100 lb. of sugar used at the start) is also added. The whole mass is then boiled for four to five minutes and allowed to stand for 24 hours. On the third day, the strength of the syrup is raised to 68 Brix and the mass boiled for four to five minutes. The fruit is allowed to remain in the syrup for three to four days. Finally the syrup strength is raised to 70 Brix. When the preserve cools, it is packed in dry glass or stone jars.

Preplanting application of herbicides Pre-planting applications are treatments made to the soil before the crop is planted. Incorporation of the chemical into the soil is essential for obtaining good results. Some examples of compounds failing in this group are triazine herbicides (*e.g.* atrazine, propazine and simazine), the oarbamates and thiocarbamates (*e.g.* propham, chloropham, eptam, tillam, avadex, CDEX), the urea-substituted herbicides (*e.g.* diuron linuron), the benzoic acid group (*e.g.* amiben), CDA, Sesone, TCA, endothal and trifluralin.

Presowing application of manure Cow dung, compost, oil cakes, molasses and other organic manures are applied to the soil before sowing.

Press mud This is originated as an effluent from the sugar mills. It contains double the quantity of nitrogen as compared to dung manure. This should be put in the field 2–3 months before sowing of the crop, it should not be used in a quantity of more than 100 maunds per acre. It contains 0.8–1 per cent nitrogen, 2.5–3 per cent phosphorus and 0.8–1 per cent potash.

Preventive measures of weed control Weed seeds germinate earlier, the seedlings grow faster, flower earlier, run to seed in profusion and mature ahead of the crop. Weed seeds have the tenacity to persist in soil or in water for years without losing their viability. Another characteristic of weed seeds is the possession by them of the phenomenon of dormancy which is the intrinsic physiological power of the seed to resist germination even under unfavourable conditions. These various mechanisms are helpful in the wide distribution of weed seeds. The abettors in the dispersal of weed seeds far and wide are water, wind and animals, including man, price fluctuations and weather.

Priming The tobacco leaves touching ground become dirty. To pluck these is called priming.

Principles of crop rotation Crops shallow rooted should follow the deep rooted crops. After leguminous crops, non-leguminous crops be sown. Crops which require less fertility should follow crops with high manure requirement. Crops which need less water should follow crops with high water requirement. Such crops which need too much of weeding and hoeing should precede the crops with less requirement of these operations. Fields should be included in the rotation which fulfil the domestic needs. Crops rotation should be such that it may be economically advantageous to the farmer. Only such crops should be included in a rotation of which the grower has much experience regarding cultivation practices, and damages and control of the insect-pests and diseases of the crops.

Process of ammonification This can briefly be shown as:

1. NH_4 ions +	Enzymic oxidation in presence of active association with CO_2+Ca-nitrate bacteria	NO_2 ions in E with bases especially Ca

2. NO_2 ions	Enzymic oxidation in presence of active association with Ca-nitrate bacterial	NO_3 ions in CO_2 E bases. especially Ca

The organisms autotropic in his case, obtain energy by the transfer, liberate some carbon dioxide, and incidentally, leaves as a by-product, nitrogen in the nitrite form.

Propagation There are two ways of producing new plants– propagation of plants. Sexual and asexual. Sexual are by seeds and asexual by means of any vegetative part of a plant.

Protein intake The availability of protein diet which induces increased quality of food by making use of fish farming etc.

Psychrometer A hygrometer for measuring atmospheric humidity in which a current of air is artificially induced in order to produce the maximum evaporation at all times at the wet-bulb thermometer, thus ensuring more reliable results. One method of producing the current of air is the use of a small fan (Assmann's psychrometer), while anoherre requires the thermometers themselves to be whirled around in the air (the sling psychrometer) *See* Hygrometer.

Psychrophilic Describing micro-organisms that can live at temperatures below 20°C. Compare Mesophilic, Thermophilic.

PTTE *See* Polytetrafluoroethene.

Pubescent Covered with soft, short hairs.

Pulses These are crops belong to the *Papilioneacious* groups of plants. Being high in their protein content, they are important for vegetarian diet. These supply vegetable proteins as essential adjuncts to predominantly starchy diets. Being leguminous crops possessing root nodules, they fix free nitrogen from the air and thereby enhance soil fertility.

Pulses for green manuring Pulses from an essential part of the vegetarian diet, and are grown commonly as pure crops in the rotation or mixed with cereals, oil-seeds and fibre crops. Roots and stubble of these pulse legumes return to the soil small quantities of organic matter rich in nitrogen. The introduction of groundnut in the cotton-jowar rotation of central southern and western India, the growing of a quick manuring variety of mung (*Phaseolus aureus*) before wheat the planting of peas in the standing crop of irrigated cotton and the sowing of sanhemps in standing paddy crop or of val (*Dolichos labab*) in the coastal paddy areas are valuable practices in soil improvement. All these leguminous crops leave the soil in better physical condition and richer in nitrogen content. Growing of pulses mixed with cereals all over India; the intercropping of cotton with groundnut or tur (*Cajanus indicus*) in central and southern parts of the country; or with cluster beans, mung and *Phaseolus aconitifolius*, or wheat with peas and gram and of fodder sorghum with *Dolichos labab* also enrich the soil.

Pulse grains Grams and tur are the principal pulse crops, occupying between them nearly half of the total acreage devoted to pulses.

Pummelo *Citrus grandis*. Fam. Rutaceae.

Pumps for lifting of water These are three types of pumps: (*i*) piston to plunger pump, (*ii*) centrifugal pump, and (*iii*) turbine pump. All the three types work on suction principals. All of them are force pumps *i.e.* not only they lift water but also raise it to a height.

Pure line In plants, pure lines are obtained by setting, which halves the heterozygosity each generation, while in animals inbreeding tends to increase homozygosity.

Pure line seed Percentage of pure germination seed; determined by multiplying the pure seed percentage by its own germination percentage and dividing the product by 100.

Pure seed As per international rules, pure seed shall refer to an include all varieties of each species under consideration as stated by the sender or found by laboratory test. In addition to the pure and damaged seed of each kind or variety under consideration, pure seed also includes undersized and shrivelled seeds as well as broken seeds of more than one-half of their original size. Seeds of Leguminoseae, Cruciferae and Coniferae with the seed coat entirely removed are regarded as inert matter. Free caryopses of grasses and cereals removed from the glumes, lemmas and paleas are considered pure seed.

Purity analysis equipment The primary purpose of equipment used in making a physical purity analysis of seed is to help the analyst to do efficient and accurate work.

Purity analysis of seeds The purity analysis place should have abundant natural non-glare light. It is preferable to locate the windows in this section along the north side of the building. Where such a location is not possible, special sun-breakers should be provided on the windows. It is also desirable for the bottom window pane along the seed purity area to open in a horizontal manner so that the air coming through the window will be deflected upward and not below directly across the working table of the seed analyst.

Purity test A determination of the four percentage purity items–crop seeds, other crop seeds, weed seeds, inert matters, based on examining quantity specified under rules.

Purity work boards These boards made of wood, locally made, provides a clean, comfortable work area for doing the purity analysis. Special purity separation tools and forceps are used for separating the inert matters, other crop seeds, weeds, etc.

Putrefaction The breakdown of proteins when decomposed by the bacteria, emit certain bad smelling substances like methane, hydrogen sulphide and organic acids are produced and this process is called putrefaction. They return nitrates to the soils and nitrogen and oxygen to the air.

PVC *See* Polychloroethene.

Q

Qualitative character A character in which variation is discontinuous.

Quality control over fruit and vegetable products With a view to exercising quality control over fruit and vegetable products, an Order called the Fruit Products Order, 1955 was promulgated by the Government of India under the Essential Commodities Act. Specifications of the various products have been clearly laid down therein and no product which does not conform to the prescribed specifications can be sold in the market. The preservatives and colours that may be used in the products have also been clearly laid down. The order also indicates the specific hygienic and sanitary measures to be adopted by the manufactures. No one is permitted to manufacture fruit and vegetable products without previously obtaining a license from the Agricultural Marketing Adviser and conforming to the requirements of factory hygiene and sanitation.

Quality control in the seed processing plant After management techniques and processing have been upgraded to improve quality and maintain identify of seed, a quality control system in the processing plant is easy to develop and operate. Such a system should help identify problems affecting seed quality, how they can be corrected in current lots and how they can be prevented in future lots.

Quality of farm manure It is also improved by the concentrated feeds given to the cattle. Cotton seed, cotton seed cake, licensed meal, wheat bran, grain husk, groundnut cake, Bengal gram, horse gram etc. are rich in nitrogen, phosphorus, potassium, magnesium and sulphur. It has been found that in the case of adult working cattle, about 80 per cent of nitrogen and the other mineral elements contained in the feed is recovered in urine, faeces and other animal by-products. Accordingly, manure from cattle feed on cereal straws hay is much less valuable than that from animals fed on legume hays, grains and concentrates.

Quality of a good sub-stratum for seed germination (1) It must be easy to handle and use, (2) It must have good water holding capacity, (3) It must be cheap, (4) The colour should make a good background for judging seedlings, (5) It must be non-toxic, and (6) It must be free from pathogens and if it is to be used again must be easily sterilized.

Quality of irrigation water Irrigation water carries a good deal of silt and organic debris in suspension or salts in solution or both according to the source or supply. River and tank waters carry material in suspension as well as in solution, whereas well waters usually contain only dissolved salts. The nature and quality of materials so carried influences the soil as well as the crop irrigated.

Quality of produce In order to ensure to the farmer a fair return commensurate with the quality of his produce, the government of India passed the Agricultural Produce (Grading and Marketing) Act in 1937. The Act provides for grading and marking of agricultural and other produce.

Quamquat fortunella Fam. Rutaceae *F. japonica* (round fruits), *F. Margarita* (oval fruits).

Quantitative variation (continuous variations) A common form of variation in which a character has an average value and members of the same population or species show various gradations from this value. Examples are height or skin colour in man. It is the result of the action of many genes (polygenes) each of small effect.

Quantitative character A character in which variation is continuous making separation into distinct classes difficult.

Quantity of organic manure The quantity of manure to be applied to unirrigated crops varies from three to five cartloads per acre receiving less than 30 inches of rainfall to eight or ten cartloads in areas of heavier rainfall. If sufficient farmyard manure is not available, it may be applied at the usual rate to a part of the land, say one-third or one-fourth the area, in rotation every year, so that all parts of the field receive manure regularly once in three or four years. For irrigated field crops, the rate varies from 10 to 20 cartloads. Sugarcane, maize and garden crops like potatoes, turmeric, ginger, vegetables and fruits receive still higher doses, amounting sometimes to 40 to 50 cartloads. A cartload of manure, measuring 30 cubic feet, weights about half a ton.

Quarantines While resistant varieties of crops offer an effective means of disease control, there are other efficient methods by which these hazards can be avoided. The selecting and breeding of resistant varieties are time-consuming processes while these alternative methods can be more quickly applied. The first among these is quarantine. *See* Destructive Insects and Pest Act 1914.

Quartile deviation Semi interquartile range of quartile deviation – 1/2 (Upper quartile–Lower quartile).

Quartz A soil mineral. SiO_2, it usually predominates sand. Quartz is silicon dioxide and is found in all crystalline rocks, also occurs as sand and is the principal component of granite and sandstones.

Quince *Cyndonia oblonga* (vulgaris), Fam. Rosaceae.

Quincunx system This system of planting is the same as the square. In this way, the number of trees is almost doubled but the distance between the central and comer trees is greatly reduced.

R

Rainfed farming Farming dependent only on rainfall.

Radial drainage A drainage pattern in which streams radiate from a central peak or upland mass. Dome structures commonly develop radial drainage.

Raffia *See* Shield budding.

Raising seedlings It is better if the plants to be grafted later on should be raised in pots in case of mango where enarching is the common practice. In case of Citrus, seedlings may be grown in the field beds itself. Other crop seedlings may be grown as per the use of the seedling.

Rasins A dried product of grapes. Most rasins are dried in the sun, often by spreading them in thin layers on trays in the vineyards. Occasionally evaporators with artificial heat are used when there is danger of rain.

Rambutan *Nephelium lappaceum*, Fam. Sapindeceae.

Randomization (Sta) It is necessary for an objective comparison between treatments to fulfil one more conditions, *viz.* the random allocation of the treatments to various plots. It can be shown that the statistical procedures employed in making comparisons between treatments will hold good only provided the treatments are allocated randomly to various plots.

Randomized block design (Sta) One of very common use in field experiments and trials is this design. The design is of wider applicability and several treatments can be tried together the same way.

Random sample The selection of a representative subset of the total population being studied by a method that ensures that each member of the population has as much chance of being selected as part of the sample as any other. This makes it possible to infer statistically that the data used is valid for the total population. One method is to use random number tables to select the sample. *See* also sampling.

Range (Sta) The range of a distribution is the difference between the largest and the smallest of observations and gives us some idea of the amount of variability present.

Raspberries-Black, *Rubus occidentalis, Rubus lasiocarpus*. Fam. Rosaceae.

Raspberries-Mauritius, *Rubus rosaefoliu*. Fam. Rosaceae.

Raspberries-Mysore, *Rubus albecense*. Fam. Rosaceae.

Raspberries-Red, *Rubus idaeus*. Fam. Rosaceae.

Ratoon In pineapple shoots which grow in the axils of the leaves are called suckers. Those which come out below the surface of the soil are called suckers or ratoons. Shoots borne on the fruiting stem, below the fruits are known as slips. Stump and slices are also planted in pineapple. In sugarcane when the canes are harvested above ground, the remaining stumps with roots are left as such in the field to be grown as the next season's crops. This is known as ratoon cropping.

Readily available portion of farm manure Only a part of the nutrient constituents of farm manure is readily available. In general, only about one-half of nitrogen, one-sixth of phosphoric acid and a little more than one-half of the potash are readily available to plants.

Reclamation of waste lands Act for this kind of land waste has been passed which makes it possible for the government to declare any area as Reclamation area and acquire it for the purpose of reclamation and improvement.

Reclamation The making of flooded derelict, or waste land suitable for agriculture and other uses. For example, the Dutch have reclaimed large areas once beneath the Zuider Zee for agriculture and settlement. Parts of the lower Swansea Valley, until recently an area of waste tips and ruined buildings, have been made into a recreation park. Areas of sandy heathland in W. Denmark have been reclaimed for farming by the heavy use of fertilizers.

Recommendation In International Seed Testing Association Rules, this means that it is only recommended and not mandatory that the procedure be followed as given.

Records of kind payments An appreciable portion of farm produce is utilized in making payments for services obtained for farm operations. A complete of such payments record should be available with the farmer for determining the total production of each crop, pattern of expenditure, cost components, desired mode of payments and farm income and expenditure. This kind of record may be maintained thus:

Month	*Name of Commodity*	*Purpose for which Payment was made*	*Quantity*	*Value*

Rectangular system In this system, the trees are planted in straight rows running at right angles and is a very neat system, easy to understand and easy to layout. Some form of the rectangular plan is used in a majority of cases all over the world. If the distance from tree to tree in the row is the same as from row to row, this is square system. This allows cultivation and irrigation in two directions, is most popular among all plants. If trees are planted closer and rows are left with more space in between, it is rectangular system.

Recycling The reusing of material or products that have been discarded because they are no longer of value for their original use. Recycling is increasing in importance as some resources become scarce. Glass, metal and paper are the most common materials that are recycled.

Reducer bacteria Under imperfect drainage and aeration, at least some of them are facultative and may readily abstract oxygen from nitrates and their reduced products. The trend of the reaction may be represented as:

NO_3	→	NO_2	→	NO	→	NH_2OH	→	H
Nitrate		Nitrite		Hypo-nitrite		Hydroxylamine		Gaseous nitrogen

Registered seed The progeny of foundation seed normally grown to produce certified seed.

Reheading in this operation, the long branches are cut back so that they do not attain undesirable lengths. Because when they bear fruits, they are borne on height out of reach and also the long branches are more likely to break when laden with fruits.

Reinforcement of farm manure Sometimes the manure contains an unbalanced amount of nutrients. In order to obtain most economical results with manures, liberal applications of phosphorus are used in connection with it. Sometimes, however, part of the phosphatic fertilizer is mixed with the manure before the application is made. It is this practice that is spoken of as reinforcement.

Relative humidity The ratio, expressed as a percentage, between the amount of water vapour actually present in an air mass and the maximum amount that the air mass could hold at that temperature. Saturated air has a relative humidity of 100 per cent. It varies inversely with temperature, given a fixed amount of water, vapour, decreasing as the temperature increases and increasing as the temperature decreases. Relative humidity cannot be measured directly, but is calculated by applying the readings from a hygrometer to a set of pre-prepared tables.

Rejuvenation *See* Neophosis.

Relation between mean, median and mode An interesting method of estimating approximately the mode is by the formula: Mean–Mode–3 (Mean–Median).

Relative measure A ratio which is independent of units. A coefficient is often, but not always a relative measure.

Replicated progeny row (Sta) These methods are used in plant breeding to switchover from mass selection to progeny breeding.

Replication (Sta) The repetition of the treatments under investigation is known as replication. Since the variation in fertility cannot be allowed directly owing to its unpredictable nature, the experimenter seeks to average out its influence over the different treatments by replications.

Requirements for germination test of seeds Seeds require certain conditions for normal germination in laboratory. These are substrata; moisture, temperature and light.

Requisites of randomization Selection of more conspicuous members in a population should not be heeded to because this will not obtain a value approaching the population mean; that is the relation E(m)=u will not hold good. If one tries to guard against selection of too many plants with one character, one may err in the opposite direction and choose too many plants with the same character. Moreover, over estimation of the mean is not always the only object of sampling; variability is also important as in genetic studies. A random sample alone can reflect the different characters of the population within the limitation of its size. Personal selection can never be trusted to give such samples.

Resistance block for water determination It consists of a block made of an inert porous material such as gypsum, fibreglass, ceramic or nylon in which two electrodes are embedded. The block is buried in the soil and the wire leads from the electrodes are connected to a specially designed meter.

Resistance *See* immune.

Resistant rootstocks It often happens that certain varieties or species which are otherwise desirable, are susceptible to insect or disease, while others may be entirely or largely immune. If the pest is one which attacks only the roots, it can be overcome by the resistant varieties of rootstocks.

Respiration Gaseous interchange between an organism and its surrounding medium; normally the taking in of oxygen and the giving out of carbon dioxide.

Resolution of velocities and forces This process is reciprocal of composition of the velocities and forces. When we find any two components of a velocity or a force at 90° to each other, they are called rectangular components.

Rest period in temperate fruits This period is found in temperate fruits like apple, apricot, and the like which go in dormant stage in low temperate and shed their leaves in autumn season. They produce new growth in the spring season.

Retardation of movement When the velocity decreases, the acceleration is called retardation.

Retting It is the process of separating the fibre of flax from the stem. This is brought about by the combined action of the bacterial (*Bacillus subtilis*) and a species of Mucor. They secrete pectinase enzyme which dissolve the pectric meddle lamellae of the cell wall.

Rhizobium Seeds of pulse crops if treated with Rhizobium bacteria culture before sowing in the field, yields of the grain is increased. This genus of bacteria of which leguminosorum species is important because they fix the atmospheric nitrogen in the root-nodules of leguminous plants. They help naturally in providing nitrogenous fertilizer to the crop and soil.

Rhodes grass *Chloris gayana*. This is a perennial grass. It can tolerate drought to considerable extent and is also winter-hardy. This grass can be established by seeds, or vegetatively by rooted cuttings.

Rice (*Oryza sativa* L) Rice is the staple food of the people world over. It is a crop of very wide physiological adaptability, being grown in both tropical and temperate conditions, from sea-level to about 7,000 feet above sea level, and from the semiarid tracts to the very wet areas of nearly 3,000 varieties, differing in morphological and agricultural characters, are in cultivation. The varieties differ in their season of growth, maturation period, suitability to different conditions of soil, rainfall, temperature, altitude and adaptability to such special environments as flooded land, alkalinity and salinity, acidity of soil, depth of standing water etc. Consistent with such wide adaptability, the methods of cultivation practised in the country differ a great deal as well.

Ridger plough These have double mould boards. There are of three sizes, the light, the medium and the heavy. The heavy ones are used for cultivation of sugar cane and the lighter ones for cultivation of vegetables and other crops. Some of the improved ridger ploughs have arrangement for adjusting the width of the furrow by changing the distance between the two mould boards. For vegetable gardens, orchards and for farmers who grow sugarcane, potatoes, etc. the possession of the ridger is a must.

Ring budding A ring of bark ½ to ¾ of inch long and containing a good healthy bud is loosened from the scion wood and slipped off one end of the branch. Top of the stock is removed portion of bark is pealed off. This ring containing the bud is slipped down over the stock until it reaches the bark.

Ring grafting *See* bark grafting.

Ringing Same as girdling.

Ring irrigation In this system of irrigation, a conduction channel is made between two rows and this supplies water to the separate basins of the individual trees. By this water supply is limited to root-hairs and thus water is fully utilized in this way. *See* irrigation.

Ringing of trees Fruitfulness can be induced by increasing the concentration of carbohydrates in the top of the tree without markedly reducing the vegetative activity. This has obvious advantages. To achieve this, ringing or girdling is regularly practised with some fruits and occasionally with others. In its most common form, it-consists of removing a ring of bark about half an inch wide or less around the trunk of the tree or around a branch.

Rock gardening Also called Rockery. In this, such ornamental plants are grown which need only a small-thin layer of soil and also can grow on stones, pebbles, shells etc. Cacti are the most popular plants used in such conditions. These may be both flowering as well as non flowering plants.

Rogue A variation from the standard type of a variety or strain.

Rogueing Pulling out the affected plants which are diseased or off-types. Virus diseases can be controlled by ruthless eradication by rogueing and burning the rogued plants. The removal of undesirable of individual from a variety in the field.

Roll mill This separates seeds by differences in surface texture and shape of the seed. The roll mill is a finishing machine and should be used only after the seed has been carefully cleaned and separated from the chaff and trash.

Room or walk-in germinator A germinator area large enough to walk into. It may be a complete room or a portion of a room.

Root crop A plant that is cultivated mainly for its root, which may be used as food for humans or animals. For example, turnips, carrots, parsnips and swedes are eaten by both humans and animals. Marigolds are fed to animals while sugar beet is processed to extract the sugar content, the residue providing cattle feed.

Root crop vegetables This group includes carrot (family Umbeliferac), Radish and turnip (Cruciferac) and beet-root (Chenopodiaceae). They thrive best in cold season but are tolerant to high temperature, especially radish. The edible portion is the enlarged root, which is rich in carbohydrates and contains minerals and vitamins. The carrot is especially rich in carotene which gets converted into vitamin A.

Rooted stipes Parts of old rootstocks used as new planting material.

Root grafting The best method is to lift the seedlings with naked roots. Pot them and later report them with about three inches of the pot. The scion is then enarched into the pot.

Root pruning A regular pruning of roots is harmful for the trees. If at all needed, it should be done only in rainy season. If root pruning had to be done before six weeks before blossoming season. If there is no rain, the pruned trees must be irrigated. This operation is practised commonly in Citrus, guava and Litchi.

Root pruning of guavas Also called exposure of roots because they have a poor fruit bearing. With holding of water and exposure of the roots with or without actual pruning of roots can also give good results. It is essential in Litchi where roots cleaned of the aerial roots give good results.

Root propagation Covers a number of methods in which a complete plant is grown. Most common forms of rootage are stem cuttings and layering, in both of which growth of roots from portion of the stem is induced.

Rootstock A vertical, usually short underground stem. It is found in many angiosperms, *e.g.* rhubarb and strawberry and in certain pteridophytes *e.g.* Osmunda. It can reduce vegetatively. *See* also vegetative propagation. In vegetative propagation of plants by graftage is included the various forms of budding and grafting in which a part of the stem of one plant known as scion is joined to a rooted plant or root itself. This is known as rootstock in such as way that the scion grows to form the fruit bearing part of the tree in future.

Rootstock scion relationship This relationship is important because a mutual influence is generally observed in the tree in view of the structure and stature of the tree form, time or fruit maturity and size, hardiness against extreme weather conditions, disease and insect pest resistance, yield of fruit, precocity, quality of fruits, vigour, longevity of the tree. This relation is between rootstock and scion but vice versa is not studied so far.

Root suckers It is a method of propagation. If lateral roots are severed from the tree or wounded, two or three feet from the trunk, they will in most cases send up shoots which may then be transferred and transplanted to the nursery as new plants. This is common for guava.

Root vegetables Such plants of which underground parts are used for vegetables such as radish, carrot etc.

Rotary cultivators Amongst rotary cultivations, small types carried on two wheels and operated by 5 to 8 H.P. engines are likely to prove useful for small holdings and market gardens. They produce to combine effect of ploughing and cultivating. Heavier rotary hoes work up to ten inches deep and need about 6 H.P. per foot-width on average soils. Some efficient types of small tillers have appeared in foreign markets suitable for rice cultivation.

Rotation of crops A systematic succession of different crops on a piece of land in order to avoid soil exhaustion. Crops are rotated in a cycle that usually varies between two and eight years. Rotation also enables weeds, diseases and pests to be controlled more effectively. Nitrogen-fixing plants, such as clover or beans, normally follow wheat or barley, which use up large amounts of nitrogen.

Rotenone It is the native drug found in a number of plants in India and extracted mainly from cube and derris roots. It is primarily a contact insecide, but is used to some extent as a stomach poison also. It is not dangerous to man or the larger animals. Its use seems to be increasing, perhaps because of legislation against the more dangerous poisons.

Rothamstead lysimeter This is one of the best types of Iysimeters of the type where small column of soil is entirely isolated by appropriate means from the land surrounding it. *See* lysimeter also.

Rotting of green manure plants For the proper rotting of the green manure, it is necessary that the green material should be succulent and there should be adequate moisture in the soil. Plants, say at the flowering stage, contain the greatest bulk of succulent organic matter with a low carbon: nitrogen ratio. The incorporation of green manure crop in the soil at this stage allows quick liberation of nitrogen in available form, with advancing age, the percentage of carbonaceous matter in plants increases and that of nitrogen decreases. If material with a wide carbon; nitrogen ratio is ploughed under, micro-organisms bringing about its decomposition draw upon released nitrogen and mineral nutrients and use a temporary nutrient deficiency.

Round-hole screen of seed processing The size of a round hole screen is the diameter of the perforation. For example a $1^1/_8$ screen has round perforations 1/18-inch in diameter. A common numbering system lists perforations larger than the size 5½ (5½/64) in 64th of an inch. Openings smaller than this size are listed in fraction of an inch; the next smaller size is ½.

Routing a sample through the seed laboratory If a seed laboratory is to operate efficiently, a system must be followed, the routing has been found to be successful.

Rye A hardy cereal grass grown on the poorer soils of N. and E. Europe and the N. central States of the USA. The grain is used to make a form of whisky and is also milled to produce a dark brown flour that is used for making bread. The grain can also be fed to livestock.

Row placement *See* band placement.

S

Saccharum officianarum **(sugarcane)** It suffers from many diseases including the smuts. Smut is caused by *Ustilago scitaminea*. It is the biggest source of cane-sugar in the world.

Saddle grafting It is the method sometimes used, particularly with plants with thick fleshy tissues, such as papaya. The top of the stock is cut from both sides to form a wedge and a corresponding cut is made in the scion so that it will fit tightly over the stock. It is bound in this position and may be waxed. *See* wedge grafting.

Safflower (*Carthamus tinctorius* L.) Safflower is said to have originated in the mountainous regions of Abyssinia and Afghanistan. At present, it is cultivated in India, Egypt, China, Persia, Southern Europe and some parts of United States of America.

The crop is cultivated primarily for its seeds which yield oil, though at one time it is used to be grown for the extraction of dye also. The seeds are edible and eaten after roasting. Their oil content varies from 24 to 36 per cent depending on variety, soil, climatic and other conditions. The cold pressed oil is golden yellow in colour and is used for culinary purposes or adulterating ghee and sesame oil or for the manufacture of soap. The oil obtained by hot dry distillation is black and sticky and is used only for greasing well ropes and leather goods exposed to weather. Safflower oil has also good drying properties and is, therefore, used in the manufacture of paints, varnishes and linoleum. It can be mixed with white paint without any after-yellowing effect. The cake, particularly from decorticated seed is used as a concentrated cattle feed, and that from undecoriticated seed is sometimes used as manure.

Saffron Kashmire farmers have a monopoly of growing this crop.

Sakiyeh A primitive irrigation device for lifting water from a channel or river onto land at a higher level. It consists of a wooden water wheel with buckets instead of paddles attached to a spindle and cog wheels. The wheel is turned by a draught animal such as a donkey or water buffalo, which follows a circular path round the wheel. Water raised in the buckets pours out as the wheel turns round and round.

Sample receiving and preparation for seed testing Samples are received by the receptionist who makes the necessary entries in the register and codifies them so as not to reveal the identity of the sample. The samples are then passed ahead.

Sampling The technique that is used to obtain a subsets of data that is representative of the total population or total area that is under study.

Sand bench A table with sides to hold sand approximately four to six inches deep for germinating seeds.

Sanitation Of orchards. Just as sanitary habits are important in maintaining human health, orchard sanitation is necessary as a basis for fighting orchard ills. Weeds harbour both diseases and insects and should not be allowed to grow in the orchard at least for long periods. Most important of all is the maintenance of vigour by means of proper cultivation, irrigation and manuring.

Sannhemp (*Crotalária juncea* L.) Sannhemp or Bombay hemp is an important crop grown for bast fibre or for green manuring purposes. The fibre is light coloured, somewhat coarse, strong, twines, nets and rough cloth for grain bags. The cellulose is used for making wrapping paper and plant stalks, and leaves are fed to cattle as green or dry fodder.

Sapindaceae soapberry family Litchi belongs to this family. Four members are in the family Nepheliceae. The Litchi and long an being subtropical with arils free and the rambutan and pulasan being tropical with avils adhering to the seed.

Sapota (*Achras Sapota* L.) Known popularly as chiku in western and central India, is adaptable to a large variety of conditions of soil and climate. It withstands the winter of the Punjab and western Uttar Pradesh, flourishes in the heavy rainfall areas of western and southern India, and grows equally well in the drier parts of peninsular India. It is an evergreen, growing and flowering almost throughout the year. Rain or cloudy weather does not harm the setting of fruits. Young plants are easily injured by frost, but grown up trees can withstand frost of a short duration. Its soil requirements are not very exacting, but grows best in alluvial or sandy loam soils having good drainage.

Scalper Seed lots often come from fields with such a high percentage of trash that it is difficult to move them through elevators and even more difficult to process them properly. The scalper is designated to remove the bulk of this trash from seed to facilitate elevating and processing operations. Using scalper on trashy seed lots before they enter the processing line will remove trash that would clog elevators. Prevent clogging of feed and increase capacity of cleaning machines and thereby speed up the processing operation.

Scarifier huller Many seeds such as lucerne, berseem, sweet clover and lespedeza, contain 'hard' seed which cannot absorb moisture and do not germinate within a reasonable time after they are planted. To obtain uniform germination and an even stand, it is necessary to scratch the hard seed coat so that moisture can penetrate into the seed. This process is known as scarification.

Scateur *See* Hand shear.

Scion In a plant graft, the part that is grafted on to the stock. The scion may be bud or shoot. *See* graft.

Screen selection This is a most important factor affecting the performance of the air screen cleaner for seed processing. Over 200 sizes of screen perforations are available and any combinations of these may be used. The openings in the perforation metal screens can be square or rectangular. *See* also selection of screen.

Sectorial chimaera A plant with a deeply seated sector or genetically distinct tissue.

Sedentary agriculture The farming of land at a fixed location instead of migrating from one site to another. Nearly all the world's agricultural systems are sedentary, the two main exceptions being shifting cultivation and nomadic herding.

Seed This is a basic input in new agronomy developing these days. The pace of progress in the field production would largely depend upon the seed with which one is able to multiply and market seeds of high quality. It is the ripened ovule, consisting of the embryo and coats, with often some additional stored food. In application, seed (beej) is different from grain (dana).

Seedling The embryo or young plant from the time it emerges from the seed until it is entirely dependent on food manufactured by itself. It consists of an apicotyl, one or two cotyledons, hypocotyl and root; the single cotyledon is usually held within the seed coat in monocots.

Seed blowers A mechanical device to remove light weight material by blowing. It is easier and easier than hand picking.

Seed certification A means to maintain and make available to the public sources of high quality seeds and propagating materials of superior varieties so grown and distributed as to insure genetic identity. Seed certification does this by means of inspections of fields and seeds, and by regulations for checking on the production, harvesting and cleaning of each lot of seed.

Seed control Seed law enforcement based on law which applies to all seeds sold commercially. Correct labeling with respect to the kind of seed, germination, purity and other quality factor is of primary importance.

Seed dressing fungicides Seeds are treated before sowing with chemicals as in case of wheat and barley loose smut and also covered smut by thirum, vitavex carbonis, Raxil have recommended for smut control.

Seed drill Quite a large number of improved seed drills both automatic as well as hand operated have been designed. Amongst the hand operated ones, a copy of the wooden tifan has been made. It is made of a mild steel with arrangement for changing the distance between the lines.

Seed drills and planters The seed drills can sow all kinds of small seeds like wheat, oats, barley, jowar and paddy, can be adjusted for variations in size of seed, depth of seeding, seed rate and the distance between the rows. Draw by wheel type or light crawler tractors can sow upto about 30 acres in a day depending on soil conditions.

Seed generation (S.G.) S.G. is seed directly produced by cross-pollination and from which F_2 plants will grow; S.G.2 represents self bred seed from F_1; and S.G.3 is self-bred seed from F_2 etc.

Seed germination for testing The preparation of samples for a germination test and the evaluation of seedlings is completed in this room, which may be called as a germination room.

Seed law The primary objective of the seed law and the seed testing laboratories is to help assure, the cultivator of a continuous supply of good quality planting seed.

Seed lot A uniformly blended quantity of seed designated by a proper number or mark.

Seed processing It is that fragment of the seed industry responsible for upgrading the quality of seed by removing foreign material other than seed and undesirable seed, improving the planting condition of seed and applying chemical protectants to the seed. Some of the services seed processors perform for farmers are (1) make possible more uniform planting rates by proper sizing and removing seed appendages which hinder planting (2) improve seed crop marketing by improving quality and maintaining dependable standards for planting seed (3) prevent the spread of seeds by removing weed seeds from crop seed (4) improve crop quality by removing seeds of other crops from the pure seed (5) protect crops from insect and diseases by applying chemical protectants (6) reduce seed losses by removing high moisture foreign material and by drying seed which are too high in moisture and (7) facilitate uniform marketing by providing storage from harvest time until the seed is needed for planting.

Seed rate To get high yields, the seeds of all the crops should be sown as per prescribed rate of seed leading to a proper plant population in the field.

Seed sampling records Information submitted with the sample should include (1) date samples, (2) who sampled seed, (3) name and address of sender, (4) kind and hybrid or variety, (5) origin or class of seed, (6) sample designation or lot number, (7) quality of seed in lot, (8) kind of tests desired; germination, purity or moisture, or all.

Seed sample drawing equipment Effective seed sampling can be done by taking the sample by hand. However, various triers and probes are used for sampling. Long triers-64 inches (150 centimetres) are used for sampling heaps and godowns. The 35 inches (84 cms) are best suited for sampling cereals and seeds of similar size from bags of 40 kgs–75 kgs capacity stacked in the horizontal position. The smaller 24 inches triers (58 cms) are ideal for sampling small free flowing seeds those which do not stick together.

Seed testing In modern times, seed selection, seed extraction, processing and storage have been established on a scientific basis instead of on mere intuition and tradition. A scientific basis has also been evolved on the systematic testing of seed for the various factors which control the quality of seed.

Seed testing laboratories The main object of seed testing laboratories is to obtain accurate and reproducible results regarding the purity, composition, the moisture content, the rate of occurrence of weed seeds and the percentage of seeds that can be expected to produce normal seedlings under favourable conditions. In some instances, such additional information as the presence of seed borne fungi, varietal purity is desired.

Seed treaters The application of seed treatment materials is usually an operation which is required at different stages. For this purpose, some machines are used as to do the work more thoroughly, elaborately and correctly. Many seed treaters are available that can apply a small amount of chemical and spread it uniformly over the surface of each seed. These vary in size and capacity for use.

Seepage drains Seepage drains are controlling the ground water table. Open drains, tiles and molas are used for this purpose. Tiles are being tried on experimental basis, as the economy of installation is to be justified at the present stage. Open drains are being increasingly used.

Sehima nervosum A good fodder grass.

***Sehima dichanthium* grass cover** The characteristic perennial grasses are *Sehima mervosum, Dichanthium annulatum* (Palwan), *Crysopogon montanus, Dichanthium caricosum, Iseilema laxum wighti, Ischeamum aristatum, Heteropogon contorous* and *Eromopogona fovelatus*. The annual species are represented by *Heteropogon contortus, Andropogon pumilus, Thermeda* spp., *Ischaemum rugosum, Aristida* spp., *Gracilea rovaleana* and *Tragus recemosus*. Most of the perennial grass species, in particular *Sehima nervosum, Dichanthium annulatum* and *Iseilema* spp. are highly palatable and nutritious. Predominance of *S. nervosum* on gravely soils and that of the latter two on level black soils will form the best management. Under severe grazing, the grassland deteriorates, resulting eventually in an increase in the stand of *Aristide* and other annual grasses. Important leguminous associates of the cover are *Atylosis scarbasoides. Heylandia latabrosea, Desmodium diffusum* and *Alvsicarpus* spp.

Selection of screen for seed processing In cleaning round shaped seed, select a round holed top screen and a slotted hole bottom screen. The round hole top screen will drop the round seed through the smallest possible opening, and scalp off everything larger than the good seed. The slotted hole bottom screen holds up the round good seed but drops out split and broken seed, small weed and small inert material. Thus the smallest possible perforation is used for scalpin, while the largest possible perforation is used for grading. To clean elongated seed, use a slotted top screen and a slotted bottom screen. It may be necessary to use a round hole bottom screen to affect a special separation, but usually better cleaning is accomplished when both top and bottom screens are slotted. For lens shaped seed, better results are obtained by using a slotted hole top screen and round hole bottom screen, but will lie flat and pass over a very large round hole bottom screen.

Selection of fruit trees For the fruit; healthy, disease free and without insect infestation and vigorously growing plants should be selected for planting in the orchard. It is also a method of selecting the desired plants from a progeny of crosses and is commonly used as a plant breeding technique.

Selection of site for growing vegetables The land selected for vegetable cultivation should have a loose, friable free-draining soil which does not easily get waterlogged. A loam or sandy loam is usually preferred, though a clay loam may be suitable if there is no impervious layer within about three feet from the surface. A stiff, hard clay should be avoided. The site should not be over-shadowed by trees or buildings. It should have good irrigation facilities.

Selection of site for wells The most important consideration is the presence of a satisfactory aquifer or water bearing stratum. Valleys transversed by rivers or irrigated by a network of canals are ideal sites as seepage and percolation from rivers and canals help to recuperate the wells. In locating a site, a knowledge of the surrounding area, its geological formation and botanical composition help considerably to minimise the risk involved in sinking a well.

Selective cropping for dry farming Rabi jowar should receive two hoeing in October–November or November–December, depending on the date of sowing. The crop of deeper soils may receive an additional hoeing, particularly if there are heavy late rains which encourages weed growth.

Selective crops against weeds Some crops are better competitors than others. For instance, crops like Sudan grass, soybeans and sorghums are good competitors while flax is poor. Raising of highly competitive crops (also known as smother crops) is useful in reducing weed infestation. One disadvantage with many of other smother crops is their slow seedling growth. If these are grown for seed, some measures have necessarily to be taken to control weeds in the early growth of these crops. However, if they are meant for fodder, the thick sowing will naturally shade out the weeds.

Selective herbicides Foliage application. These treatments are made to the leaves of growing weed plants, usually as sprays out in a few cases as dust application.

Selective weedicides Such as 2,4-D are being used. When this weedicide is sprayed in between rows of crops, it kills the weeds only and not the crop. Sprayers are used for spraying these weedicides.

Selective logging The periodic cutting down of mature trees or those growing too close together in a forest to allow optimum use to be made of the forest resources. This most commonly practised where a forest consists of a variety of trees of different ages.

Selective role of nodule bacteria While the nodule bacteria are considered to be common to all leguminous plants, it is well known that the organisms from one species of legumes are not necessary well adapted to the ready production of tubercles on other leguminous species although infection many be affected by proper culturing. Grouping for this purpose have already been identified.

Self-sterility The condition in many hermaphrodite animals and plants whereby male gametes cannot fertilize female from the same individual. *See* incompatibility.

Separation Used in vegetative propagation which means involvement of only of removing from the parent plant of a part which would grow into a complete parent without man's intervention. Also called Division.

Separation of seed in process A seed processor should approach each lot of seed as methodically and analytically as the research scientist approaches his work. The processor should first carefully examine the seed to determine what his processing problem will be, and then decide how can solve the problem to produce a final lot of the maximum purity and germination.

Sericulture The rearing of silkworms to produce raw silk. This is a labour intensive form of farming using cheap skilled workers. Fresh mulberry leaves to the silkworm the thread unravelled from the cocoon.

Service sample of seed These are those samples submitted by a cultivator, seedman, cooperative or seed farm. They have no official status but must be tested promptly to provide the sender immediately information about the quality of seed with which he is dealing.

Sesame (*Sesamum indicum* 1.0) Sesame or sesamum is said to be a native of India, though sometimes its origin is traced to Africa and sometimes to the East Indies. It is an important oil seed crop. The oil content of the seed generally varies from 46 to 52 per cent. The oil percentage extracted is about 40 to 42 per cent in the expellers and the rotary mills and about 36 to 38 per cent in the village ghanis. The seeds may be eaten fried, mixed with sugar or in the form of sweetmeats. Sesamum oil is used as cooking oil in South India. It is also used for anointing the body, the manufacture of perfumery oils and for medicinal purposes. Oil cake is edible and eaten avidly by working class. It is also a valuable nutritious food for milch cattle.

Sesone-sodium 2, 4-dichlorophenoxyethyl sulphate, a herbicide.

Set of simple tools for farmers Simple set of tools are recommended for every farmer using agricultural machinery. Wood saw, small drill, hammer, three sixes of chisels, a wooden plank, rough files, auger, pinches, steel file, hacksaw pipe, wrench, adjustable wrench, a one foot piece of girder to be used as an anvil.

Setaria italica A small millet.

Setaria sphacelata A very promising fodder grass.

Setts The internodes in sugarcane stem are known as setts. These form the planting material for a sugarcane crop.

Sexual propagation *See* propagation.

Sewage manure The liquid wastes like sludge and sewage contain large quantities of plant nutrients and are used for the growing of sugarcane, vegetables and fodder crops near many large towns by operating sewage farms. In many places, the undiluted sludge has been found to be top strong for healthy plant growth and if it contains readily oxidisable organic matter, it actually reduces nitrates present in the soil. The disadvantages are still great if sewage is used on land without a preliminary treatment.

Shaddock Second name for Pummelo.

Shallow wells Shallow wells are usually hand-dug. The diameter varies with the depth and quantity of water expected. Those wells that are situated near irrigation tanks or similar water reservoirs contain mainly percolation water. They are usually very shallow and supply water for irrigation, sometimes for many months even after the tanks dry up. In the case of surface or open wells, the diffing has to be continued until a good spring or water-bearing stratum is reached. If the well is sunk in soft earth, the sides are built up to prevent them from caving in. No such precaution is necessary, except near the top, where a well is sunk in soft earth, the sides are built up to prevent them from caving in.

Shape and size of blocks and plots (Sta) In field experiments, the error mean measures the soil fertility variation within blocks. This meant square and hence the standard error of treatment comparisons, therefore depends upon uniformity of land chosen for experimentation as well as the shape and size of blocks and plots.

Sherbats (Syrups) Syrups of rose, khus, sandal, kewra, etc. are prepared by boiling sugar and water with a little citric acid to avoid crystallization of sugar. Generally 6½ lb. of sugar, 3½ lb. of water and ¼ oz. of citric acid are mixed together and boiled for 10 to 15 minutes or until the syrup becomes thick. They are then cooled, desired flavour and colour added and filled in clean dry bottles.

Shield budding Also called T-budding and is most common among budding. This is very commonly done in roses and citrus. The scion wood should be mature and rounded but not old. It is desirable but not necessary that leaves still should be present. The blade of the leaf should be cut off, leaving the petiole as a handle. A vertical cut is made just through the bark 1½ inches long on the stock. At top of this, a horizontal cut about half an inch long is made. If necessary, the bark is then loosened from the wood using the back on the blade. The bud is then cut by passing the blade of the knife under it, so as to remove a shield shaped piece having a bud on it about 1 to 1½ inches long just slight wider than the bud. This will include a small amount of the wood inside which may or may not be removed. The bud now should be immediately inserted under the bark of the stock in the T-shaped cut. It should then be wrapped in order to hold it firmly in position and to exclude air and water. This may be done with waxed-tape string, raffia or even banana fiber.

Shield Budding

Shifting cultivation An agricultural system in which a patch of forest is cleared of trees and most vegetation, to be cultivated for a few years until the fertility of the soil is seriously reduced. The site is then abandoned and another is cleared elsewhere. Cleared vegetation is usually burned (slash and burn) and crops planted in the fertile ashes. Certain cash crops are also grown by shifting cultivation method; when yields fall and the land is abandoned and a fresh site is cleared.

Sibs Progeny of the same parents derived from different gametes.

Sib mating The mating of sibs.

Side dressing In this method of manure application, manure is placed between rows or on all sides of plants.

Side grafting It is done without removing the top of root stock which may be considered larger than the scion. The base of the scion is cut to form a wedge with one side slightly longer than the other. An incision is made in the side of the rootstock at an angle of about 20 degrees, deep enough so that the end of the scion may be inserted. Tying may not be necessary but waxing is.

Significance of large samples It is the ratio of deviation and standard error, hence formula:

$$\frac{\text{Deviation}}{\text{Standard error}}$$

Assuming that the sample means are distributed normally, which assumption is usually reasonable even when the original population deviates somewhat from the normal, the above ratio may be regarded as a normal deviation. By reference to the table of probability integral, it can be found that the probability (1-a), that the deviate is numerically as large as or larger than the above got value, between x and y, these values of probability corresponding to values as above is found to be z etc. etc.

Silvex 2(2,4,56-trichlorophenoxy) propionic acid. A herbicide.

Silviculture The cultivation of trees including the theoretical and practical aspects of forestry. In the UK, the Forestry Commission is responsible for the establishment and maintenance of forest land that is not privately owned. Great emphasis has been put on quick-growing conifers to produce large quantities of softwood.

Simazine 2-chloro-4, 6-bis (ethylamin)-s-triazine, a herbicide.

Simazine in maize This chemical as such is harmful to all plants but its use as an herbicide in maize is due to the possession of the mechanism in maize which converts it to the non-toxic hydroxysimazine.

Simple forms of nutrients in soil These are ammonium salts, nitrite salts, nitrate salts, phosphates of calcium, potassium etc., potassium salts such as sulphates and carbonates.

Single cross A cross between two inbred lines.

Slips *See* ratooms in pineapple.

Slopes on screen Same as pitch of screen.

Slope and strip cropping Very little experimental work has been done to determine the best combination of crops and other factors in a strip cropping programme.

Slope	*Width of Erosion Permitting Crops*	*Width of Erosion Resistant Crops*
1 per cent and below	150 feet	30 feet
Between 1 and 2 per cent	80 feet	20 feet
Between 2 and 3 per cent	45 feet	15 feet

Sludge *See* sewage.

Sewage proper The mechanical clogging by colloidal matter in the sewage and development of anaerobic organisms which not only reduce nitrates already present in the soil but also produce alkalinity. Bacterial contamination makes the eating of raw vegetables grown on untreated sewage or sludge a real danger to health. For these reasons, it is now usual to provide a settling or a septic tank in which sewage is allowed to stand to relieve it of the heavier portion of the solid matter in it or undergo a preliminary fermentation.

Small nut (*Chiraunjl*) *Buchananis latifolia*. Fam. Anacardiaceae.

Small tractors A small tractor can be defined as a machine four wheels or two axles. Smoking Food stuffs like pork and fish are subjected to the action of smoke for their preservation. The preservation is done by the chemical action of gases produced along with the smoke.

Sod soil It is in which weeds and other plants have been included during cultivation. Such operations are not useful under Indian conditions and tropical cultivations also.

Sod Under certain conditions, fruit may be grown without cultivation. In the cooler moist regions of Europe and America, trees are often grown under sod. Grass is planted or the natural grass is allowed to grow, beneath and below the trees of the soil is not stirred in any way. The grass is moved from time to time and preferably left on the ground. This practice is in contrast with the method known as clean cultivation, which is increasingly popular and in which the soil is cultivated and the growth of weeds is prevented at least during part of the year. A cover crop or green manure crop may be grown during part of the year. The effects of sod seem to be mainly reduction in soil moisture and available nitrogen.

Soil application by fumigation Soil sterilants fall in this group. They are used where it is desired to remove all plant growth or to keep areas free of plant growth and is mainly meant for weed control.

Soil application of herbicides These treatments are made to the soil before emergence of the crop (pre-emergence), or after emergence of the crop (post-emergence). These have to be carried into the soil by rainfall, overhead irrigation or a sub-irrigation that wets the treated area. Selectivity is dependent upon plant tolerance, location of herbicide in the soil and difference in the growth habit of crop weeds.

Soils and crops for dry farming The soil characteristics like depth, permeability, water retention capacity, configuration of land etc are some of the factors that affect the dry farming practices. The alluvial soils in the dry farming regions of north are fairly level, deep, light to medium in texture with favourable physical characteristics and good permeability. Thus, even small showers of rain are useful and there is utilisation of most of the water held by the soil due to low moisture content at wilting point. The greater depth of these soils is the most potential factor.

Soil horizon Well defined layer within the soil profile, parallel to the local ground surface. Three main horizons are characteristic (A, B and C) of many free draining soils.

Solanum tuberosum It suffers from many fungal and virus disease which are very destructive and sometimes prove to be a challenge to grow this crop. Potato is an important vegetable crop.

Solanum xanthocarpum Fam. Solanaceae. A perennial weed herb. 2,4,5-T 1 lb./acre a.e./100 gallons per acre at young stage before leaves become tough.

Solder An alloy used in joining metals. The molten solder alloys with (wets) the surfaces to be joined, without melting them, and solidifies on cooling to form a hard joint. The surface must be clean and free of oxide and a flux is often used to achieve this. Soft solders consist of lead with upto 60 per cent tin and melt in the range 183-250°C. Soft soldering is used, for example, in plumbing and making electrical contacts. Brazing solders are copper-zinc alloys that have higher melting points and produce stronger joints than soft solders; silver can be added to produce silver solders.

Soncchus oleraceus Fam. Compositeae. To control, apply 2,4-D volatile easter (low) 1.5 lb. ae. At early bloom stage or aminotriazole (4 lb. in 100 gallons of water) plus wetting agent (6 fl. Ozs/100 gallons spray at flowering stage).

Sorghum (*Sorghum vulgare* pers.) Next to rice, sorghum or jowar has the largest area among the Indian food crops. It is the staple food of the people in the relatively dry tracts. Grain may be broken and cooked in the same way as rice. It can also be ground and the flour used for a variety of preparations, being inferior only to wheat for bread making. Some varieties are good for the making of popped grains. Jawar grain can also be malted to manufacture food for infants and invalids. The plant stalks, green or dry, provide nutritious fodder for cattle. Usually, the stalks of the grain crop are cut into lengths of one foot or 18 inches before being used as cattle food.

Sorghum halepense Fam. Gramineae. A perennial weed. Control same as for *E. crusgalli.*

Sorting fruits Object of sorting is to remove over-ripe or damaged fruits, which is likely to decay before reaching the customer, often also damaging good fruit in contact with it to remove as culls such fruits as is so green or so deformed as to lower the standard of the whole lot. Such culls may sometimes be sold locally at a low price or used in making by-products. It is obviously to the advantage of all concerned that such fruits be separated as early as possible.

Source of ammonium sulphate $(NH_4)_2$ SO_4 It is got to some extent as a by-product from coke and gas and also synthesised.

Source of carbon for plants *See* photosynthesis.

Sources of irrigation water There are two main sources of irrigation water: surface water and ground water. There are water supply projects, such as reservoirs, flumes, ditches and dams, water measurement and control; methods of applying to land; the amounts adequate for different crops; with this there is a need for maintaining irrigation works for their efficient use.

Sour orange Family Rutaceae *Citrus aurantium.*

Sour soap *Annona muricata,* Fam. Anonaceae.

Sources of fruits From the reference to different fruits in the writings of various authors in different countries, it appears that these sources may be of three kinds–(1) indigenous, (2) those introduced from other parts of Asia and Europe before the great age of exploration which started late in 15th century, and (3) those introduced later. This does not mean that it is easy to assign each fruit to one of these groups to any confidence as in many cases our present knowledge allows only a tentative classification.

Sowing depth or seeding depth Seeds should be sown at proper depth to get good germination. The depth at which seed should be placed in soil for its proper germination. Same as seeding.

Sowing devices Amongst the indigenous implements, many seed sowing devices have been employed on implements such as the desi plough, commonly called nari plough, tifao, and other multi-type implments that are used in Mysore state. Tube or tubes are fixed to these implements and on the top of these a seeding bowl, in which the seed dropped is mixed. The seeding bowl has corresponding holes in it. The construction of the hole is indigenous so that the seed is fairly evenly dropped into the tubes and down to the tines. In some areas considered backward, agriculturally, broadcasting or sowing of seed at random on the cultivated field is also followed, but major portion is now-a-days sown in lines or drilled. Sowing in lines ensures proper distribution and replacing of seeds, uniform, spacing of the plant and economy in seed rate. It also helps regular germination ensuring better crop stand and facilities for carrying out inter cultivation by means of bullock operated hoes, thus reducing the cultivation cost.

Soya bean A plant that is widely cultivated for its seeds. These are planted annually and the plant is used as a green manure or a forage. Also the seeds have widespread uses; they contain about 35 per cent protein and are eaten whole as flour, and as soy sauce. Soya-bean oil is used in wide variety of products including margarine, paints and cooking oils. The residue is used as fodder for livestock and as a meat substitute for man. As a leguminous crop, the plant adds nitrogen to the soil. It is grown in USA, China, Japan and now also in India.

Spacing of trees Distance to be allowed between trees depends on a balance between several factors. Most authorities agree that most orchards are generally planted too closely for the best results but also disagree as to the space that should be allowed. The maximum development of the individual tree requires that it is planted where it will come into competition with no other tree. The roots of a tree extend much further than do the branches, several times as far under certain conditions.

Spa A place that is noted for the medicinal properties of its mineral springs (named after the Spa, a watering place near Liege in Belgium).

Spawn The mushroom spawn used in growing the fungus, consists of a block of richly manured soil which is permeated with it.

Special manures This is the name given to such inorganic manures which supply only one nutrient element.

Speed of screen shake It is adjusted to control the action of the seed of the screen in processing. Fast shake causes seed to turn and tumble and present all their sides to the screen openings. Fast shake speeds are, therefore, very effective in cleaning chaffy seeds. Fast shaking also helps more grass seed over wire mesh screens.

Speed sprayers The modified sprayers developed recently.

Spinifex (porcupine grass) A tufted very spiny dense-growing that grows in Australian deserts. Spinex is drought resistant and often develops on sand dunes, which it stabilizes.

Split-plot and strip-plot designs (Sta) That is a highly modified design for precision and accuracy of the tests of many factors.

Split doze application of manures By giving manure to crops bit by bit at different times is best utilised by the crops and there is minimum loss of elements by this method. It proves economical as well as there is not possibility of some kind of special chemical reactions taking in the soil. The manure applied is mostly utilised and in case of less moisture present in the soil, there is no possibility of the manure being wasted and the growth of the crop is uniform.

Spoilage Of fruits and their products. Caused by the microscopic organisms or fungi including moulds, yeasts and bacteria. The most common moulds are those belonging to *Penicillum* and *Aspergillus*.

Spray or nutritive elemental solution If the deficiency of nutritive elements cannot be ascertained, sprays in solution form can be used for this purpose. In this method, surface and plants are sprayed with nutrient solutions. Sprays of iron, copper and zinc are likely to damage the leaves so as to avoid this their solutions are made in combination with a small quantity of slaked lime.

Spreaders Many spray materials do not adhere well to the foliage or fruit especially when the surface is waxy. This also tend to form large drops rather than a thin casting over the surface. Materials which help to overcome these tendencies are called spreaders and stickers. Soap is an excellent spreader for nicotine sprays but should not be used with all materials. Casein is also effective and may be used in several forms, the most common being calcium casienate. Fresh and dried skimmed milk is also satisfactory. Blood albumin is now commonly used. Rasin is a good sticker. Flour, especially if it has a high gluten content is very satisfactory helping both spreading and sticking and at the same time having no chemical reaction with sprays.

Sprinkler method of irrigation *See* aerial methods of irrigation.

Spring typed harrow The spring-typed harrows are lighter implements weighing about 25 lb. per type. They can work upto a depth of five inches and cover about one-third of an acre per foot-width per hour.

Sprout Put forth new shoots of plants.

Spurs *See* scale leaves.

Squashes and barley waters Oranges and grape fruits are peeled and juice extracted with a screw-type juice extractor; lime juice is extracted with lime squeezers; pineapples are peeled, sliced and juice extracted in a basket press; mangoes are peeled and pulp extracted by screening through a sieve. From juicy mangoes, the juice is squeezed by hand. The juice is then mixed with sugar, citric acid etc. For barley water, barely flour is made into paste and boiled with a little water before.

Square system of planting In this system after leaving about 7 metres space all around the field to prevent any effect of hedges or windbreaks, straight lines are ploughed in the field. If the space is not desired to be left, other alternative would be to dig trench all round in the space between the hedge rows and the rows of trees. Then rows are drawn in the field with the layout as between tree to tree. And in this way for this layout, the pits are dug for planting the trees of fruit plants. This system is simplest of all and helps in doing various cultural operations with much convenience.

Stages between harvested seed to marketable seed *See* figure.

Standard deviation (Sta) This measure of dispersion is calculated by squaring the deviation of each observation from the mean, adding the squares, dividing by the number of observations and extracting the square root.

Standard deviation of the mean (Sta) It is commonly termed the standard error. If X stands for the standard deviation of means of samples of n individuals and Q for the standard deviation of the parent population.

Standing crop The nutritional portion of the biomass in the area at a given moment.

Stages Between Harvested and Marketable Seed

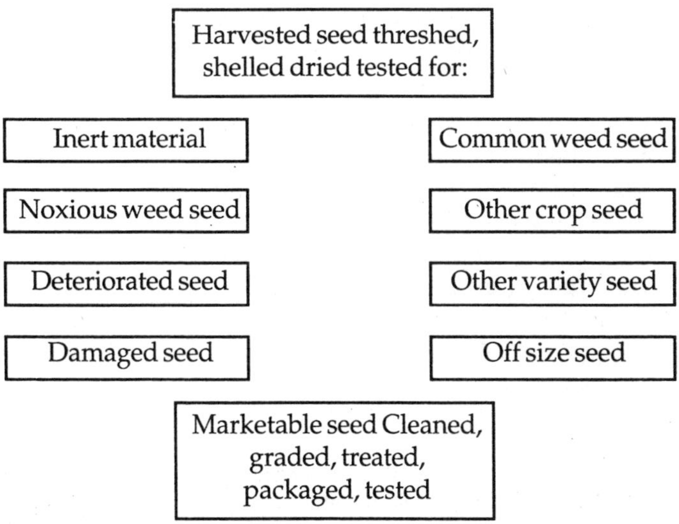

Staple (1) Principal item that is produced, consumed or traded in an area. (2) The length of quality of a textile fibre. Short-staple cotton, which is rather coarse, is grown in India whereas finer cotton with a long staple of 50 mm, is grown in Egypt and USA.

Staple length The hair (fuzz) attached to the cotton seed do not have the same length. But even then some of them are uniformly of same size. The average length of these staples is called length.

Starter In making of compost, starter has an important role to play. It is generally mixed in the pit of organic manures so the carbon-nitrogen ratio in the pit be at a suitable stage and this is necessary because only under proper ratio into the bacteria and fungi which play an important role in rotting of organic material of the manure material. When there is a deficiency of nitrogen in the manure, then it is advisable to add a nitrogenous starter in the manure pit.

Starter fertilizer solution This manure is applied mostly in the transplanted crops. Immediately after transplanting plants are normally irrigated but in this case, plants are irrigated with a dilute solution of the manure. This helps in quick establishment of the newly plant crop and helps them to grow quickly.

Steamed bone 1-2 per cent nitrogen and 22–30 per cent phosphorus.

Stem-galls of coriander This disease of coriander (*Coriandrum sativum*) caused by *Protomyces macrosporus* appears in the form of tumour-like swellings on the veins, leftstalks, peduncles and other green parts of the plant.

Stevension screen The standard shelter for meteorological instruments at a weather station. It consists of a white painted wooden box with levered sides and a double roof the base of the box being above the ground. It allows the true shade temperature to be measured by allowing free circulation of the air, while protecting the instruments from such outside influences as direct sunlight, radiation from the ground and buildings, rainfall and strong winds. At all weather stations, it contains a hygrograph and thermograph.

Stickers *See* spreaders.

Stone drains They are also of some value in certain cases, they must always be short and are likely to clog. Also their drainage is slow and insufficient.

Storage It is an exercise of human foresight by means of which commodities are protected from deterioration and surplus supply are carried over the future consumption in the season of scaricity.

Storage of fruits For getting good price of the produce, the grower may need proper storage conditions. This should avoid spoilage and injury to the fruit in storage. Also *See* Cold Storage.

Storage and warehousing Proper storage of goods and produces plays a very important part in the development of 'orderly' marketing. It helps to reduce the cost of handling and movement, ensures safety against loss by depredations of insect pests and minimises the scope for dishonesty in dealings. It helps small agriculturists to hold up their stocks till prices improve. It also provides an acceptable form of security on the basis of which credit can be obtained from banks on reasonable terms.

Strawberry (*Frageria vesca* and *F. chiloensis*) Fam. Rosaceae. The cultivated varieties of strawberry grown in India are also imported. The following are recommended: Laxton's Latest, Royal Sovereign, Early Cambridge and Huxley Giant. Maiden plants (runners) that have not borne any crop are used for planting. Planting distance is half a metre between plants, three-fourth of one metre between rows. Runners with good root system are utilised to set a new plantation. Transplanted in March-April in the hills and in January-February in the plains of India.

Strength of a mixed manure This is quoted in three numerals such as 4 : 2 : 5. This means that this manure has the proportion of nitrogen: phosphorus: potash as 4 : 2 : 5 respectively.

Strength of a staple It is measured by different methods. Weights are hanged on the count and individual staples and their capacity to bear the weight is observed. Weak staple is likely to break with small weight even but a strong staple can endure more weight. A 4 counts staple should be strong enough to lift a weight of 5 lbs., though some has a capacity to hold 5 to 6 lbs. also. It has also been noted that in comparison to wool, the cotton staple is 3 times more strong than wool in view of their length.

Striga asiatica Family Scrophulariaceae. A perennial root parasite. 2, 4-0 2 lb./acre in 80 gallons of water 2-3 weeks after seeding of host; fenac 3 lbs/acre in 50 gallons of water, pre plant soil incorporation especially in sugarcane.

Striga euphradoides Same as *S. asiatica*. Also called cane killing weed of sugarcane.

Strike-off stick A wooden stick received with the measure weight tested to be used when leveling the seed in the volume measure.

Strip cultivation (strip cropping) The growing of different crops on parallel narrow strips of ground, usually following the contour pattern of the land. This type of cultivation reduces the risk of soil erosion particularly if, on sloping ground, cover crops such as clover are alternated with crops that leave some of the soil exposed *e.g.* maize.

Strip cropping for dry farming As a general rule, strip cropping should be practised. Erosion-permitting crops like jowar and bajra should be grown in strips separately by those of erosion-resistant crops, such as ground nut, hulga and matki, which have plants with a spreading habit. For effective protection, the cover crops other than groundnut, should be sown at two to three times the normal seed rate. A part of the eroded land should be utilised for establishing an annual grass cover or planting improved grasses on contours.

Stub grafting A scion about 6" long is cut in wedge shape and inserted in a slanting incision on the upper side of the side shoot to be replaced, which is pressed down to allow the scion to be inserted and released to hold it fast. The shoot is then cut off just above the point of union and the wound sealed with grafting wax.

Submitted sample A sample that has been sent to a seed testing laboratory for testing.

Subsistence agriculture (subsistence farming) A type of farming in which output is consumed almost by the farmer and his family leaving only a small proportion for sale. Where there is a small surplus, it is sold at a local market so that the family can buy basic provisions such as cooking pots and clothing. Subsistence farming is found in many countries of the Third World, such as India, Sri Lanka and Nigeria. The crops produced are known as subsistence crops.

Subsidiary tools for farmers Farmers should also maintain some spare screws, nuts, bolts, washers, etc. By keeping and learning to use such a set of tools and maintaining commonly required spare parts, a farmer would be able to carry out simple repairs.

Sub-strata for seed germination test Suitable sub-strata for seed germination in laboratory include paper towelling, blotter paper, filter paper, cotton, and soil. Check test should be made on each new batch of paper by comparing it with material known to be toxic.

Superphosphate A mixture–mainly calcium hydrogenphosphate and calcium sulphate used as a fertilizer. It is made from calcium phosphate and sulphuric acid. Made from phosphate having a constitution as:

$$CaH_4(PO_4)_2 \text{ and } Ca_2H_2(PO_4)_2$$

It is the most widely used phosphoric fertilizer. Prepared formerly by treating bones with sulphuric acid, it is now manufactured largely by treating ground phosphate rock with almost an equal quantity by weight of sulphuric acid. This treatment produces a brownish grey mixture containing monocalcium phosphate and calcium sulphate (gypsum) in practically equal quantities. This fertilizer is manufactured in three grades; single superphosphate containing 16 to 20 per cent phosphoric acid; dicalcium phosphate 35 to 38 per cent and triple superphosphate 44 to 49 per cent.

Support price As a long term measure, the government guarantees the farmer that in the event of glut in the market, it would not allow to price falls below economic level, hence fixed minimum support prices for essential farm products.

Suckering *See* topping and sucking.

Sucker A underground shoot that at some stage emerges above the soil surface and give rise to a new plant, which initially is nourished by the parent plant until it becomes established. Suckers are troublesome in certain ornamentals (*e.g.* roses) in which the plant has been grafted onto a wild root stock, single sucker shoots develop from the root stock and grow at the expenses of the ornamental. Also *See* ratoon be liasiana.

Suction pressure (SP) *See* osmosis.

Sudan grass *Sorghum sudanensis*. A high yielding, draught resistant grass, it has proved to be a valuable grass as an acclimatised forage grass. Like *Andropogon sorghum*, it develops hydrocynic acid in the early stages and should therefore be harvested and fed only in the flowering stage. This leafy fodder compared well with *Andropogon sorghum* and is relished by all kinds of livestock.

Sugar beet A plant with a green foliage and large carrot-shaped white root, which grows best in moist mild climates. It is cultivated for the sugar content of its roots, which can be over 20 per cent by weight. The juice is extracted and treated to produce sugar crystals; the green tops provide fodder and the dried pulp is also fed to livestock.

Sugarcane (*Saccharum officinarum* L.) Sugarcane is the important source of sugar and holds a pre-eminent position as a cash crop. A tall bamboo-like grass from which sugar is obtained. The grass is mainly grown on plantations in tropical and subtropical regions where there is a long wet season followed by a cool dry season, during which the cane ripen and develops about 10–15 per cent sucrose content. The canes, 3–8 m high, are cut and crushed between rollers to obtain sugar solution. This is refined to produce sugar crystals; a side product of molasses, is used as an animal feed stuff and is fermented to produce rum.

Sugarcane crushers Growers of sugarcane cannot dispose of their entire harvest as cane. Part of it has to be converted into gur or jaggery. The gur is also a popular edible food in the villages. Formerly the juice was extracted from the cane by crude two-roller wooden crushers. With the advent of growing of improved varieties of cane, a three roller cast iron crusher operated either by bullock power or by engine power has been put into the market. This has got greater crushing efficiency and, therefore, farmers are either buying such customers or taking them on rent during the season.

Sulphurus acid It is good preservative with 0.2 per cent solution.

Summer crops For growing these, the climate is hot and humid.

Summer season annuals These are Amaranthus, Balsam, Celosia, Cock's comb, Coreopsis, Gillardia, Gompherana, Marigold Portulaca, Potania, Sun flower, Salvia, Turenia, Varbisinia, Zinnia Ipomea spp.

Summer weeds These weeds appear only during summer season and infest the summer season crops.

Sunflower Belongs to Family *Compositeae, Halianthus annuals*. Very rich in edible oil content.

Super-cooling The process whereby water can be cooled well below freezing temperature (0°C) and yet remain in the liquid form.

Sweet lime Family *Rutaceae, Citrus paradisi.*

Sweet orange (*Citrus sinensis* Osbeck) The sweet orange is grown under sub-tropical and tropical conditions. Dry and arid conditions coupled with distinct summer and winter seasons having low rainfall are most favourable for the growth of sweet oranges. Rainfall seems to be unimportant if irrigation is provided but atmospheric humidity exerts a great influence. It can be grown on a wide-range of soils, from heavy clays to very light sands with pH ranging from 6.0 to 8.0. The tree is particularly sensitive to high concentrations of salts and cannot stand waterlogging.

Symptoms of nutrient deficiency In plants, nitrogen deficiency is shown by the pale colour leaves, the lower leaves dry and look fired. Growth of plants is retarded due to which they become small in size. The yellowing starts from the lower leaves while the upper leaves are still green. Due to major deficiency, the leaves may attain a brown colour after which they die. In grasses, these symptoms first start with the tips of leaves.

In the absence or lack of phosphorus, plants remain small in size and their colour becomes dark green because in the absence of this element, more nitrogen accumulates on the upper parts though sometimes leaves may have violet or red colour. Lower leaves start drying. Crop has a delayed maturity. Deficiency of potassium is indicated by showing the burning or firing of the tips and margins of the lower old leaves.

Synchronous flowering A crop when sown late, but it flowers at the same time as the right time sown crop.

Synthetic variety A variety produced by crossing inter se number of genotypes selected for good combining ability in all possible hybrid combinations with subsequent maintenance of the variety by open pollination.

Syrup While fruit can be canned without the addition of sugar, the quality of the product is ordinarily much better if sugar is added. Sugar is generally added as syrup, more acid fruits require more sugar. The best grades of canned fruits receive syrup from 40 to 60 per cent sugar. Fruit syrups are generally made with sugar, but some of excellent quality are made by concentrating the natural juice.

T

Tables used in statistical experiments (Sta) (1) One digit random numbers (2) Two digit random numbers (3) Three digit random numbers (4) Values of normal deviation (5) for chosen levels of P (5) values of to for $P = 0.5$ and $P = 0.01$ (6) Values for F (Variance ratio) 5 per cent points of e^{2z} (7) Values of F (Variance ratio) 1 per cent points of e^{2} (8) Values of X^2 probability.

Tall armed and protected hedges *Duranta plumerii, Lawsonia alba, Murraya exotica, Polyalthia longifolia* and *Thevetica nerii* are the plants of this category.

Tangelo A hybrid of tangering and the pumelo or grapefruit.

Tangors Hybrid between the mandarin and sweet orange include one known as the Umatilla which resembles the tangelos, as well as some which are commonly considered mandarins.

Tank A cistern-like feature, water-storage pool or reservoir, particularly one used for small-scale irrigation such as those in India and Sri Lanka. Recent models of river system often consider that stream flow is generated by a series of reservoirs or tanks that store a certain amount of water before they overflow. For example, the soil cover may be considered as one tank, which can store a certain amount of water. Such schemes are called tank models.

Tapioca (*Manihot utilissima* Pohl) Tapioca is an important subsidiary food crop, possessing large edible roots, rich in starch.

Tariffs and trade Tariffs are duties levied on imports or exports, while trade (domestic or foreign) consists of the movement of commodities and services from one area to another.

Tartaric acid Best for fruit jelly making.

T-budding *See* shield budding.

Tea (*Camellia thea* link) The dried leaves and shoots from an evergreen shrub (*Camellia sinensis*), from which the beverage is prepared. The shrub grows best on hill sites where rainfall is about 1300 mm and temperatures are above 22°C during the growing season. A number of pickings by hand takes place during the growing season so a large labour force is required. In Assam (India) and Sri Lanka, the chief exporters of tea, it is mainly grown on large plantations where in China and Japan, it is grown on small holdings. Two main types of tea are produced; Indian tea, which is allowed to ferment and China (or 'green') tea, which is only partially fermented. It is also grown in other parts of SE Asia, S. Africa and part of S. America. Tea is an evergreen bush cultivated on plantation scale for its tender leaves which are processed into the cured product that, infusion in boiled water, yields the well known beverage bearing the same name. The commercial tea is either green (unfermented) or black (fermented). The leaf is rich in carrein and tannin. Tea is believed to be indigenous to the hills of northern China; however, the cultivation of tea on a large scale is only about 100 years old. Two major groups are recognised under the species, the Chinese teas, typified by var. *bohea* and the Assam teas by var. *assamica*. The former is a tall plant with soft leaves and can grow at elevations under 5,000 feet, while the latter is a dwarf bush with leathery leaves and can be grown upto 8,000 feet above sea level. A year after planting, tea when the plant is about 18 inches high, 'centring' is done *i.e.* the main stem is cut a few inches above ground. The new laterals developed by the plant are again cut a little higher up. This process is repeated. In four or five years, the plant becomes a mature bush of 18 to 24 inches height and is ready to yield a crop. In Assam, pruning is done in December–March after allowing the bush to grow more.

Temperate One of the original Greek divisions of the world based temperature lying between the torrid and frigid zones and meaning an area where there are no extremes of temperatures.

Temperate fruits The fruits which are grown in the temperate zones of the world.

Temperate and alpine grasslands Characteristic grass species of the cover between 7,500 to 10,000 ft. are *Koelaria cristate* and *Poa pratensis* associated with species of the genera *calamogrostis, Angrostis danthonia, Helictotrichon* and *Muhlenbergia*, Dominant temperature. An index of the degree of heat of a body, substance or medium such as the atmosphere. It is a fundamental component of meteorology and climatology and every weather station takes various temperature readings every day.

Thermometer An instrument for measuring temperature. Though there are several types but the most commonly used is that consists of a sealed glass capillary containing a liquid (usually mercury or alcohol) which is fed by a reservoir at the base.

Temperature alternating In germination testing, specific temperatures with seeds being held at both lower and higher temperatures for a designed time each day.

Temperature constant In germination testing, a specific temperature which should not vary by more than 1 ec.

Temporary preservation Of fruits. Proper sorting, refrigeration, use of mild antiseptics like sugar, salt and vinegar and pasteurisation constitute this method.

Tenancy reforms The measures of tendency reforms related to (1) regulation of rent (2) security of tenure (3) conferment of ownership on tenants. These points are very important economically and improvement in them will certainly yield results that go in favour of higher production and tenants' welfare.

Tension A stress produced in a body as a result of two forces pulling against each other, such forces are set up in igneous rocks during cooling, giving rise to tension joints rift valleys. Compare compression.

Tensionmeter It is an instrument consisting of porous ceramic cup with a narrow neck to which is attached a rubber tubing. The other end of the tubing is connected to a mercury manometer or to a vacuum gauge. It is filled with water and fixed in the soil at the desired depth. The instrument is based on the principle of suction force or capillary tension. The suction force or the soil moisture tension developed at the point of contact of the ceramic cup with the soil is measured by a manometer or the vacuum gauge. A movement of moisture takes place through the porous cup till an equilibrium is established between the water in the cup and the soil water. The vacuum gauge tensionmeter is better suited for irrigation control. An increase in soil water content reduces the tension and lowers the gauge reading. Conversely, a decrease in the soil water increase the tension and the gauge reading increases. With a multiple-cup tensionmeter, it is possible to determine soil moisture tensions at several depths simultaneously. The tensionmeter gives reliable indications of soil moisture at high moisture contents only. This limits its usefulness in the field.

Terrace cultivation The growing of crops on level steps or terraces that have been constructed on hillsides.

Testa Hard outer seed coat or seed coats.

Test cross It is a cross between a hybrid and a recessive homozygote.

Test of efficiency of farm labour A good test of efficiency with which labour is used are the figures of crop acreage, and gross income per worker and per head of animal. Low figures indicate surplus labour, both men and animal.

Test of independence 2 × 2 Table. Another common use of X test is in testing independence of classification in what are known as contingency tables.

Tested normal seedlings A good seed show a good well-developed root system; a well developed and intact hypocotyl without damage to the conducting tissues, an intact plumule with a well developed green leaf, within or emerging through the coleptile, or an intact epicotyl with a normal plumular bud; one cotyledons for seedlings of monocotyledons and two cotyledons for seedlings or dicotyledonous seeds.

Tests of significance In applied investigations, while comparing some characteristics such as the mean, the variance of a measure of association between two characters of a group with a specified value, or in comparing two or more groups with regard to the characteristic. In making such comparisons, it is difficult to rely on the mere numerical magnitudes of the index of comparison such as mean, variance or measure of association. Statistical science provides an objective procedure for distinguishing whether the observed difference connotes any real difference among group or can be described to mere sampling functions. Such a procedure is called a test of significance.

Testing heterogeneity (Sta) The problem of testing heterogeneity is similar to that of independence. Thus if there are several sets of data, say r, each divisible in two classes. These frequencies would be set down as a 2 x r table and the X^2 calculated as before.

Tetramethylthiurum disulphide Arasan is the common name.

Teosinte *Euchleana mexicana* A close relative of cultivated maize, is quite popular as a fodder grass. Essentially a tropical grass, it can be grown as a rainfed crop in warm and humid climate with a rainfall of over 40 inches and as an irrigated crop in dry climate.

Terai In E. India and Bangladesh, a belt of marshy ground and vegetation on the lower parts of alluvial fans. Much *terai* has been cleared for human' settlement.

Terms of trade The relative level or ratio between the price obtained for exports compared with the price obtained for imports.

Terrance A relatively flat strip of land that forms a shelf or bench. River terraces usually occur paired, one on either side of the channel and are remains of a former alluvial floodplain that has been dissected by renewed down cutting by a river. This may result from rejuvenation of the river or from a climatic change. Subsequent terrace may form steps on either side of a river.

Terrace cultivation The growing of crops on level steps or terraces that have been constructed on hillsides. Banks or wall are built to hold the soil in place and, where paddy is grown, to enable water to stand on the fields. Terrace cultivation takes place where land is scarce *e.g.* China, Indonesia, Japan or where rainfall is limited *e.g.* Italy, Greece.

***Themeda-arundinella* grass cover** Important grass species characterising the cover are *Themeda anathera, Arundinella* spp., *Eulalipsis binata, Chrysopogon montanus, Heteropogon contours, Bothriochloa pertusa, Bothriochlo intermedia* and *Pennisetum orientale*. Most of these species are of medium palatability and nutritive value. *Themeda anthera* being more palatable should be considered as the key species in the management of the cover.

Theodolite An instrument used in surveying for measuring angular distance in both the vertical and horizontal planes. It consists of a telescope that can be rotated in both the horizontal and the vertical planes. It is normally mounted on a tripod and must be leveled before use. It is extremely accurate and all readings are made by means of verniers or micrometers. It is used for triangulation surveys.

Thephrosia purpurea Fam. Leguminosae. A perennial under shrub weed. Control measures same as for *Melilotus alba.*

Thinning This is done of both the flowers and fruits growing especially in the subtropical regions. It is done to achieve a better size of fruit with attractive colour in the fruits. In apples and pears, flowers or fruits are plucked off immature to let a smaller numbers to grow, mature and ripen.

Three groups of fertilizers Based on the three main nutrient elements of a simple fertilizer, grouping can be had (1) those that of supply nitrogen (2) those serving source of available phosphoric acid and (3) those having water soluble potash.

Threshing by tractor Thresher in some of the large private as well as government farms is done by trampling under a tractor attached with a disc harrow. It has been noticed that at 15 BHP tractor with a suitable harrow produces about 4,100 to 6,560 lb. of clean undamaged grain per day of 10 hours at a working cost of Rs. 35 to 40. The bhusa or the broken straw is also of good quality.

Tidal current The horizontal movement of the water of the sea caused by the tide.

Tile drains Where the land possesses considerable natural drainage, the tiles are laid only along the depressions. This is spoken of as a natural system in that the tile facilitates the quick removal of water from the places of natural accumulation. Layout of drainage every regular system should contain of two parts–the laterals and the main drain. The laterals should always enter the main at an angle of 45 degrees.

Tile drain method of study of percolation In this, an area is chosen where the tile drain receives only the water from the land under study and where the drainage is efficient. A record of the amounts of flow throughout a term of years and analysis of the drainage water will throw light on the ordinary losses of water and plant nutrient from a normal soil under a known cropping system. The advantage of such a method is that a large area of undisturbed soil may be studied.

Tillage Tillage methods and tillage implements play a great role in brining about proper soil structure conducive to good yield. In conservation farming, tillage is essential for achieving best results. There are considerable differences of opinion about the usefulness of ploughing as preparatory tillage, particularly in dry farming areas. In a majority of investigation in such tracts in India, shallow ploughing as well as only harrowing have produced as deep ploughing. Shallow cultivation had been found adequate to kill weeds and enables oil to absorb rainfall. Fine pulverising and deep cultivation have often led to excessive soil erosion. On the other hand, in areas which have an annual rainfall over 40 inches, the good effect of 12-inch deep tractor ploughing of moderately heavy, weed infested or virgin soils are said to persist for as long as five years. Even six to seven-inch deep tractor ploughing of land already under cultivation has reduced weeding costs while increasing crop yields.

Time of fertilizer application Fertilizers applied before sowing should be broadcast uniformly and harrowed in. In the case of fertilizers containing soluble phosphates, the desirability of applying them in one to two-inch wide bands on either side or the row of seeds at a depth of four to six inches by means of a drill has been pointed out This reduces 'fixation' of soluble phosphates in the soil. It is also a good practice to mix superphosphate with farmyard manure at 40 to 50 lb. to a ton before applying the organic manure, particularly on dairy farm land. Sulphate of ammonia used as a top dressing should not be applied when plant leaves are wet.

In the case of irrigated crops, the application of fertilizers should invariably be followed by a watering. In the case of fruit trees, the fertilizer should be applied to the soil under the crown, a few feet away from the trunk. The area of application should be progressively extended as the trees grow bigger.

Time interval of green manure The time interval during which a depression of the nitrate nitrogen of the soil occurs is also influenced by the succulence of the green manure and its age of maturity.

Tin can This is made of tin plate according to desired size and is about 2 per cent tin, the base being a high quality steel, low in carbon. These cans are used in canning industry.

Tobacco damping off Caused by *Pythium de baryanum* and *Pythium aphanidermatum*.

Tobacco (*Nicotiana* spp.) In the production of tobacco, India ranks third after the USA, China of leaf produced. Of about 50 valid species of tobacco, the most extensively cultivated species are *Nicotina tabacum* and *Nicotiana rustica*. The species *tabacum*, known as *desi* and Virginia, has usually pink flowers, long and broad leaves and stout stalk. On the other hand, *rustica* commonly called *vilayati* or *kalkatia* had yellow flowers, round and puckered leaves and relatively soft (succulent) stem. Leaves of *tabacum* are used for cigarettes, cigar and cheroot, bidi, hookah, chewing and snuff purposes, while *rustica* is grown mainly for hookah, chewing and snuff.

A plant with large oval-shaped leaves that is best grown in subtropical regions but can also be grown in mid-latitudes as well as the tropics, although the quality and quantity vary according to climatic and soil differences. The leaves are dried and cured in large barns for use in cigarettes, pipe tobacco, cigars and snuff. The USA is the largest producer with North Carolina, South Carolina, Virginia and Kentucky as the main states. The other major producing regions are SE Europe, China, India, the USSR and Brazil.

Tolerances The permitted latitudes of variation between test results. They are based on statistical considerations and take into account variation in tests due to random sampling and testing.

Tolupothrix tenuis It is a most widely used algae for inoculation purposes. This form is used as it has a capacity to solublize insoluble phosphates.

Tomato chutney (sweet) Fully ripe tomatoes, after trimming all green and blemished portions, are blanched in boiling water for about five minutes and crushed. For every 10 lb. of tomatoes, 7 lb. chopped onions, 13 lb. sugar, 10 oz. salt, ¾ oz. chillies, one bottle of vinegar and a small quantity of garlic are taken. All the ingredients except vinegar are cooked to a thick consistency. The vinegar is then added and the product further cooked for four to five minutes.

Tomato ketchup Fully ripe tomatoes of deep red colour are thoroughly washed and after removing blemishes, green or yellow portions and the central core, cut into small pieces, crushed with a wooden ladle, boiled and strained by rubbing through a 1 mm sieve covered with wide mesh muslin cloth. For every six gallons of strained pulp, 15 oz. of chopped onions, 1 oz. of chopped garlic, ½ oz. of cloves, ½ oz. of mixed spices (coarsely powdered cardamom, black pepper and cumin in equal quantity), 1/8 oz. of amce (jalvatri), ¾ oz. of cinnamon and ½ oz. of red chillies are weighed out, tied loosely in a muslin bag and the bag placed in the juice to which 1¼ lb. of sugar is also added. The mass is then boiled to about one-third its original volume and the bag removed and squeezed. Thereafter, two bottles (52 oz.) of good quality vinegar, 2 oz. of glacial acetic acid, 12 oz. of salt and $2^2/_3$ lb. sugar are added and the mass boiled again till the volume of the finished product is about one-third that of original pulp. Sodium benzoate, dissolve in water, is next added at the rate of two grains per pound of finished product, by first mixing the solution with a small quantity of the finished product. The material is filled at 190°F into sterilized hot bottles which are sealed airtight with crown seal or cock, heated in boiling water from 30 minutes, stored in a cool dry place. The corked bottles may be waxed and capped with a screw-cap, if desired.

Tomato sauce This product is very popular of all the products made of tomato. It is also used everywhere as in hotels and homes equally.

Tongue incharging A tongue is formed between the cuts of the two joining parts so as to fit one's tongue into another.

Top dressing Application of manures or other chemicals supplied in a standing crop is the top dressing method. The purpose of application at such time is that it be applied only when it is needed by the crop. Nitrogenous fertilizers are generally applied in this way.

Topping To pluck out inflorescence or top new growth of plants is called topping.

Topping and suckering in tobacco Topping and suckering are important operations for improving the sizes, body and quality of leaves. In cigarette tobacco, although topping was hitherto not practised by the farmers, it is recommended and is providing beneficial except in plants with darker leaves. The crop is topped by removing the flower head after opening of about 10 per cent of the flowers. In *bidi* tobacco, about 10 to 12 leaves are retained at Nipani and about 11 to 12 leaves excluding 2 to 3 sand leaves in Gujarat area. In cigar and cheroot and chewing, tobacco in Madras retaining 12 to 14 leaves has been found to improve the yield and quality. Topping of about 14 leaves has been found to be the best for chewing tobacco in Bihar and Gujarat. In the case of hookah, tobacco is practised in both tobacum and rustica, after complete emergence of the head of inflorescence. All the leaves on the hookah plant of both the species are retained in Punjab whereas 12 to 16 leaves are retained in Bihar, UP and West Bengal. Suckering is the removal if growing shoots from leaf axils. It is done once or twice in cigarette tobacco, 8 to 10 times in bidi tobacco and 2 to 3 times in other tobaccos, except the wrapper tobacco which is not topped.

Topsoil The well developed part of the soil at or near the ground surface (A horizon). Topsoil is the most important for plant growth.

Top working When old trees are budded or grafted to some other varieties, this is done in several ways. The main branches may be cut back to within a few feet of the trunk. Of the shoots which are thus forced out, a few are selected and budded or grafted. Scions, the size of a pencil are cut wedge-shaped at the base and are inserted in the split one on each side, care being taken that one of cambium layer of the stock touches that of the scion.

This is known as cleft grafting. Sometimes the tree is cut off a foot or two above the ground. If the trunk is not more than 2 or 3 inches in diameter, cleft grafting then may be used. Otherwise it is better to use bark-grafting and scion is inserted between the bark and wood of the stock. This is also called rind or crown grafting.

Torulopsis utilis Can synthesise protein from carbohydrates and inorganic sources of nitrogen.

Town compost In recent years, large scale composing of town refuse and night soil in properly constructed trenches away from human habitation had been taken up successfully by the municipalities of many large and small towns. Trenches, three to four feet wide, two feet deep and of convenient lengths are filled with successive layers of nigh-soil town refuse and earth, in the order named. The compost gets ready in about three months.

Toxic effects of essential elements While the ten elements were recognised as essential for plant growth since a very long time, since their necessity can easily be demonstrated, the later investigations have positively added four others to the list– manganese, boron, copper and zinc. Small amounts, often mere traces, of these latter elements usually suffice, while large doses may be highly toxic under certain conditions.

Trace element An element required in trace amounts (a few parts per million of food intake) by an organism for health.

Trace elements as fertilizers Of the elements found in mere traces in soils and used sparingly by plants, the applications of four– copper, manganese, zinc and boron, under certain soil conditions and with certain crops, often gives surprisingly beneficial results.

Tractor driven implements The tractor driven implements can be classified into two parts. One set of implements is for tillage operations and another set for stationary operations. Under the tillage operations, the implements are plough, harrow, seed drill and combined harvesters and threshers. Under the stationary operations are pumps, stationary threshers and winnowers, graders, etc. In addition, the tractor can also be used for transporting fertilizers, manures or grain by attaching trailers.

Tractor ploughs The tractor ploughs are of two kinds, namely–the mould-board and the disc plough. They cut two to five furrows at a time depending on the number of complete bottoms or bodies composing the plough. In selecting a plough with the correct number of bodies in it, the maximum draw per pull of the tractor in the highest ploughing gear as well as the draught of the largest furrow normally required to be cut should be taken into consideration. Correct speed of the operation is important for good ploughing. The best speed for ploughing with norma! types of ploughs is from 1.5 m.p.h. depending upon soil and other field conditions.

Trailers The trailers are of two types. There are five-ton fixed trailers or three to five-tons self tilting trailers. Depending upon the times, the trailers are to be emptied the type of trailer should be selected.

Training grape vines These are trained in trenches. Sometimes they are trained to climb single. Other trees are trained in various ways.

Training of young trees It may frequently begin in the nursery, though some sorts are ordinarily not pruned until the time of transplanting. It is probably desirable in most cases to train the tree by pruning to a single stem at least for a short height above the ground. Some fruit trees tend to send up a number of stems, all but one of which should be removed as they appear. The formation of the main framework of the tree is the most important part of training (*See* Figures on Page 259 and 260).

Transfer of available forms Calcium and potassium held largely in mineral forms must also succumb to the solvent action of charged with carbon dioxide if they are to be reduced to simple compounds.

Translocated foliar herbicides These move within the plant, a property that makes them effective in destroying roots of perennial plants and a low volume application is possible. The physiological differences among plants determine the selectivity. Systematic herbicide is also a term employed for a herbicide which is translocated. Examples of this group of herbicides are barban, 2, 4-DB, MCPB, 2, 4, 5-T and silvex.

1: Close centre system or leader system; 2: Vase shape system; 3: Modified leader system

Transplanting machines Even though strenuous efforts are being made in different countries, such as Japan, Italy, USA and China, so far no really useful paddy transplanting machine has been prepared. Therefore, transplanting has to be done by hand. A small tool for transplanting two to three plants of paddy at a time is being used in Burma and Malaya. There is need to introduce this tool in India because with little practice, it can increase the efficiency of the labourers at least by 20 per cent.

Tree tomato *Cyphomendra beteca*, Fam. Solanaceae. An excellent natural source of colouring food products.

Trellis system Wire trellis are used sometimes but these are not very popular due to their high cost.

Triangulation A method of surveying in which an irregular area is divided up into a network of triangles. One side, the base line, is measured accurately, from which the lengths of the other sides can be computed. The triangulation provides as accurate framework for subsequent more detailed mapping as it establishes a number of initial control points.

Trials The newly evolved or introduced cultivars or varieties of crop plants are first put into test trials for observations on yield and

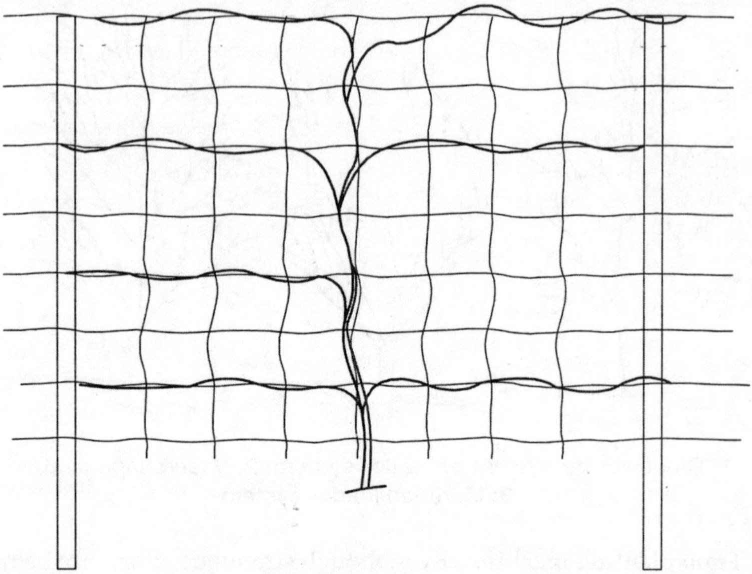

Trellis System of Training Grape Vine

its components, diseases and insect-pest resistant and finally on high productivity. Such trials should be statistically genuine and sound so that the results may be dependable for future use.

Triangle law When a particle is in equilibrium under the action of velocities or forces, then they can be represented by the three sides of a triangle taken in order.

Triangular perforations of seed processing These perforations are measured in two ways. The system most commonly used is to give the length of each side of the triangle in 64ths of an inch; *i.e.* a 9 triangle has three equal sides, each 9/64-inch long. These are identified as 9 tri, 10 tri, etc. The second system numbers openings according to the diameter in 64 ths of an inch of the largest drde that can be inscribed inside the triangle. This system is identified by the letter V following the number size, as 9V, 10V etc.

Triangular system In this system of planting, trees are planted as in the square system, except that those in the second row are midway between, instead of opposite to those in the first row. This system of planting bears the same relation to the hexagonal system as the rectangular system to the square system. Somewhat fewer trees per acre may be planted than in the square system with the rows, the same distance apart.

Trifoliate orange Fam. Rutaceae, *Poncirus trifoliata*. A good root stock for some varieties propagation as it is a hardy plant

Trisanka A promising grass also known as *Pennisetum trispecific* hybrid (*P. american* X *P. purpureum* X *P. saquomulatum*). It is a tall tussoky perennial looking similar to pearl millet and napier grass.

Trueness of type in variety testing A field plot planted with varieties from a seed control or seed certification programme or a seed-survey to check the growing plants for distinguishing morphological characteristics to verify a variety.

Tuber A swollen underground stem or root that contains stored food and acts as an organ of perennation and vegetative propagation. Stem tubers (*e.g.* potato) develop at the end of underground stems by swelling of nodes and internodes. There is an increase in pith tissue to form a round tuber that bears bud in the axils of greatly reduced leaves. The stems connecting the tuber to the parent plant then secures. Root tubers may develop in the same way from adventitious roots, as in the dahlia.

Tuber crops In recent years due to the shortage of food, the importance of growing potato and sweet-potato as subsidiary food crops has been realised and their growing on an extended scale is recommended.

The edible portions are tubers that are modified stems in case of potato and modified roots in the case of sweet potato. These types are in starch and can be kept in store for a few months under proper storage conditions.

The potato is essentially a crop of cool, moist regions whereas sweet potato is a warm season crop which requires high temperature of its best growth.

Tuber indexing In potatoes, this has helped considerably in checking disease, as it aids in the elimination of infected tubers and in obtaining healthy stocks.

Tube wells In the case of tube wells, the drilling is done by percussion tools or rotating bits or by jetting. The material loosened during the drilling operations is removed by forcing a stream or water through the tube or drill stem. After it is drilled, the well is developed by installing screens or strainers and by surging, backwashing or hard pumping to remove the fine material from the formation and to rearrange the gravel ground screens. A capacity test is usually made on every tube-well before a pump is permanently installed.

Turbine pump In a turbine pump, the impeller is built in a vertical shaft and rotates in a compact casing or bowl.

Turn-rest plough *See* mould beard plough.

TCA Trichloroacetic acid.

Tril *Trifolium alexandrium*. This plant known as *berseem*, its cultivation has become impossible because of absence of the proper bacterium in the soil to infect the roots.

Type A group of varieties of similar characteristics or an admixture of seed of indistinguishable varieties but having some characteristics in common, for example grain type jowar, yellow maize, etc. The material used to define a species. It is usually a dried specimen stored in a herbarium but may also be a drawing. The term type may also be used to describe the type species of a genus, the type genus of a family etc. For example the genus Solanumm is the type genus of the family Solanaceae.

Type of farming to be adapted This should be considered in the light of soil and climatic conditions; labour and marketing facilities. From the study of the past records, it is possible to determine the deficiencies in the system of farming and take steps to increase by making adjustments wherever and whatever required.

U

Uniform motion It is calculated as $S - u \times t$ where u is uniform velocity, t is the time and S is distance travelled.

Uniform velocity If a body covers equal distance in equal intervals of time, however, small, in the same direction in a straight line, its velocity is said to be uniform.

Uniformity of staple thickness This quality is important as uniform staple is more likely to break during ginning and weaving.

Uniformity trial (Sta) A uniformity trial consists in growing in a field piece of land and particular crop with a uniform treatment, dividing the field into small units and harvesting and recording the produce from each of these units separately.

Union In vegetative propagation of plants by graftage, the joining of scion to the root-stock is known as union.

Unit A reference value of a quantity used to express other values of the same quantity. *See* also SI units.

Unit value of manures The element present in one tonne (2240 lbs.) expressed as percentage is the unit value of the manure. Thus it can be calculated as:

$$\text{Unit Value} = \frac{\text{Value of one tonne manure}}{\text{Percentage of element}}$$

This is 22.4 times more of value of one pound.

Unoverlapping classes (Sta) The classes should not overlap. This, of course applies to the cases where observations are approximately same and differ by very small amounts.

Urea It is a white, crystalline, inorganic chemical. It is a highly concentrated nitrogenous fertilizer containing 45 to 46 per cent of non-protein organic nitrogen. It is fairly hygroscopic and presents considerable difficulty in storage, especially in humid regions. To offset this difficulty, this fertilizer is also produced in granular or pellet forms and coated with non-hygroscopic inert material. It is very soluble in water and, therefore, subject to rapid leaching. It is, however, quick acting. When applied to soil, its nitrogen is rapidly changed into ammonia. Like ammonium nitrate, urea supplies nothing of value except nitrogen. It may be applied at sowing time or as a top dressing but should not be allowed to come in contact with seed. It is suitable for most crops and can be applied to all soils. It has 46 per cent nitrogen. At locations where there is scarcity of water, it should not be used. It is used as foliar spray also. A carbamide $CO(NH_2)_2$. A white crystalline compound made from ammonia formaldehyde resins. It is the end product of metabolism in many animals and present in urine.

Urecotelic Excreting nitrogen in the form or uric acid like reptiles and birds.

Urochloa mosambicensis A good quality fodder grass.

USDA United State Department of Agriculture.

Use of ammonia by higher plants It may be synthesized into cell contents and again becomes a part of complex proteinaceous compounds, or it may appear as a simple waste or by-product ammonia.

Use of bees for pollination of crops The use of honeybees for increasing seed production and fruit crops and for improving the quality of the latter has, in recent years, been widely practised in Europe and America, but in India this still is in the theoretical stage. It has, however, been realised that bees are helpful and, in some cases, even indispensable, for crops in the manner described below. Many varieties of fruit trees cannot set fruit with their own pollen *e.g.* all almond and sweet cherry varieties and many apple, pear and plum varieties. Papaya is a fruit in which male and female flowers are borne on different trees. To enable any such tree to fruit, it is essential for it to receive pollen from some other fruit variety. The number of wild insects available, especially in early spring in the hills, is very small and ineffective for pollination, while wind pollination is negligible. Honeybees do not die out in the winter and, being under the control of man, they can be utilised for the purpose of pollination with great ease and advantage. All species of *Brassica*, which is grown for seed is fruitful only by the presence of bees and this remark also applies to clovers. Many other crops, such as pulses, lentil, cotton and oilseeds, generally benefits from bees. This is also true in regard to berries of all kinds and several vegetable crops, such as radish, turnip and cabbage when these are grown for seed.

Use of fertilizer It may be stressed that in order to secure maximum response to fertilizers, the crop should be irrigated immediately following their application and at suitable intervals thereafter. The response of the crop under arid and semi-arid conditions is usually uncertain and relatively small. It is, therefore, advantageous to restrict the use of chemical fertilizers mainly to irrigated land and areas of assured rainfall.

Use of farm manure in composting A compost is commonly made of alternate layers of manure and the vegetable matter that is to decay. Layers of soda or of soil high inorganic matter also be introduced. The manure supplies most of the decay organisms which in the presence of nitrogen, cause a rapid and effective humification.

Use of fertilizer solution in soil This method is common in USA. Anhydrous ammonia which is liquid form at a pressure of 200 lbs. per sq. inch, is pumped into the soil by certain special equipment. This prevents loss of ammonia to a great extent, and plants are not damaged in any way.

Use of geese This is for controlling weeds in cotton which is popular in some areas in South-West America. The young geese selectively feed on Johnson grass (*Sorghum halepense*) without harming the cotton. The eradication of prickly pear by the cochineal insect (*Dactylopious tomentosus* L.) in Maharashtra and Madras states is the best example of biological control in India. Claims are also made in our country of the control of *Kans* grass (*Saccharum spontaneum*) by growing basket grass, the roots of which are supposed to excrete some substance or substances inhibitory to *kanns*.

Use of random numbers It is a general property of randomly selected samples that if we give at the time of selection of the successive units of the sample each of the remaining units an equal chance of being selected. Therefore random sampling is often described as one which gives each number of the population an equal chance of being selected.

Utility of wind-breaks Wind breaks planted across the prevailing wind direction afford protection against lodging, prevent wind erosion, reduce evaporation and provide shade and timber. To be really effective, the wind-break should have a central row of tall trees, flanked on their side by rows of shorter trees and outside these, rows of shrubs and bushes, such a wind-break deflects the air currents upwards thus lessening their horizontal force. The loss of space occupied by such wind-breaks is more than made up by the timber they provide and the increased crop yields of the protected area.

Usual fertility requirement of soil Since phosphorus, organic matter and nitrogen usually need attention, especially in the production of field and vegetable crops, the system is built around their probable deficiency. In general and in dairy farming, the utilization of crop residues, farm manure, legumes and superphosphate is one of the simplest and most satisfactory combinations that can be made of the various fertility factors.

Utilization of farm labour This information is very important as manual and bullock labour account for nearly two-third to three-fourth of total expenditure on ordinary farms.

V

Vacuum counter A vacuum machine equipped with a plate containing 50 or 100 holes for automatically counting seeds for a germination test.

Value of grass in stabilization of bunds and gully control Researches conduced at Sholapur at Maharashtra have shown that the grasses have good capacity for binding the soil together. Some of these are also drought-resistant. The principal promising grasses are Blue panic (*Panicum antidotale*), Napier grass (*Pennisetum purpurium*), Rhoes grass (*Cholris gayana*), Marvel (*Dicanthium nodosum*), Pavana (*Sehima nervosum*) and Anjan (*Cenchrus ciliaris*). Similar observations made in the Nilgris Hills in Madras have indicated that *Phalaris tuberosa, Eragrostis amabilis, Eragrosits curvula* (Weeping love grass, *Cynodon dectylon* (hariali) and *Dactylis glomerata* (Cock's foot) are very effective in binding the soil, stabilizing the rises of the bench terraces and sodding water channels. In Madhya Pradesh, *Thelepogun elegans, Pennisetum pedicillatum* amongst the annual grasses and *Chloris bournei*, Thin Napier (*Pannisetum polustachyan*) amongst the perennial grasses have been found suitable for similar purposes.

Value of farmyard manure It must be stressed that the value of farmyard manure in soil improvement is due to its content of principal nutritive elements and its ability to: (*i*) improve the soil, tilth and aeration, (*ii*) increase water holding capacity of soil, and (*iii*) stimulate the activity of micro-organisms that make the plant food elements in the soil readily available to crops. The supply of organic matter, which is later converted into humus is a property of farmyard manure.

Vanada unufonica An orchard plant of commercial value.

Variability of farm manure This depends upon (1) kind of animal (2) age, condition and individuality of animals (3) food consumed (4) litre used and (5) handling and storage which the manure receives before it is spread on the land.

Variety The taxon below the subspecies level. The term is often loosely used to describe breeds of livestock or of various cultivated forms of agricultural and horticultural species. *See* also called cultivar.

A subdivision of species. An agricultural variety is a group of individuals within a species which are distinct in appearance or performance from other varieties within the same species. A subdivision of a kind which is characterised by growth, plant, fruit, seed or other characters by which it can be differentiated from other seeds of the same kind (*See* testing).

Varieties of legume vegetables Early round-seeded varieties: Asauji, Hara Bauna and Lucknow Boniya, Early Badger, Little Marvel, Early December and W.F. Messey are good wrinkled seed varieties.

Medium Bonneville, 1. 19, P. 8.

Late N.P. 29, Darantia Kaip or Talwar (Greenfeast or Lincholn), Kala Nagni or Kanawari (round), Khaparkheda. Sylvia is a good edible podded variety (peas).

Kentucy Wonder (pale), Giant Stringles Pod (bush) (Lima beans).

Pusa Naubbahar, Pusa Sadabahar, Pusa Mausmi, Sutiya Pardeshi (Cluster beans).

Pusa Barsati, Pusa Phalguni, No. 419 (Cowpeas).

Vase shaped tree The young tree is trained in such a way that leaving at 1½ to 2 ft. height branches, the whole tree upwards is cut completely. Only 3-4 branches are now allowed to grow. Next year these branches are also pruned beyond the length of 9 inches and whatever branches grow on these. Out of these leaving 2 branches all the rest are pruned away. When these remaining branches spread out, they give the young tree a shape of a cup or jar or vase (Page 259). This method of training trees is more popular because it allows more sunlight for the tree and thus fruits develop an attractive colour. *See* training of young trees.

Vegetables for health For normal maintenance of health, the daily minimum requirement of vegetables is 10 oz. per head *i.e.* 20 per cent of the daily requirement of the total food of an adult.

Vegetable bed preparation The land should be ploughed and harrowed several times to make a mellow bed. It should be more or less leveled and laid out into beds or ridges and furrows. As vegetable crops are heavy feeders, they should be liberally supplied with manures and fertilizers. An application of 10 to 20 tons of well decomposed farmyard manure or compost is very helpful to keep the soil in proper physical condition. An application of a mixture of fertilizer containing nitrogen, phosphorus and potassium further helps to hasten growth, give higher yields and improve the quantity of crop in some cases. Some crops may profitably be top-dressed with a nitrogenous fertilizer.

Vegetable growing Vegetable growing yields a much higher income than any other type of farming, a valid reason for the vegetable growing communities to make a good living out of small holdings. Vegetable crops can be produced in succession on the same plot of land, because they are short-duration crops and all the family labour of the vegetable grower can be usefully employed through out the year.

Vegetable manures Such manures which are prepared out of vegetable materials such as leaves of trees and plants, twigs and roots etc.

Vegetable nursery bed A nursery bed is usually located in a protected place. The area is ploughed about 12 inches deep and all clods broken by repeated harrowing. Well rotten farmyard manure or compost is applied at the rate of 10 to 20 tonnes per acre (about four to six weeks in advance) of sowing, and mixed well with soil. Beds are then prepared about six to nine inches higher than the rest of the field and carefully leveled. The beds are generally three to four feet in width to permit easy access to all parts of the bed. A layer of compost about one to two inches thick is spread on the surface and mixed to a good tilth. The seed is sown in rows and covered with soil. The bed is watered immediately after sowing, usually a sprinkling can have a fine spray. After the seed is sown, BHC (5 per cent) is strewn round the bed in a small band to keep ants away. The bed is kept clear by regular weeding. In order to accelerate the growth of seedling, the bed may be watered with a two per cent of solution of ammonium sulphate or nitrate of soda. The seedlings, when five to six inches tall, are ready for transplanting which is invariably done in the evening.

Vegetable oil Oil obtained from plants, usually from the seed, nut or fleshy part of the fruit. Among the seeds yielding oil are the soya bean, sunflower, cotton, linseed, maize and oilseed rape and mustard. Oil is also extracted from groundnuts, coconuts and palm kernels. The fleshy parts of the olive and oil palm fruit also provide valuable vegetable oils. Vegetable oil is used for making margarine, cooking oil, cosmetics, soap and many other things.

Vegetable seed supply One of the most essential requirements of successful vegetable growing is good seed which should have high vitality and good breeding. It should be free from foreign matter such as weed seed, dirt, etc. The seeds, when sown, should germinate rapidly giving healthy, vigorous seedling to ensure a good stand of thrifty quick growing plants. It is desirable to conduct a germination test before sowing particularly with oil seeds. A sample of 100 seeds may be sown in the field a week ahead of the sowing time, and the speed and percentage of germination should be noted. No official standards of germination for various vegetable seeds have yet been adopted in the country but a germination percentage of 70 to 80 should be considered good.

Vegetable seed procurement Efforts should be made to procure seeds from government agencies, failing which the seeds should be obtained from government approved seed merchants. It is usually cheaper to buy seed than to produce them. However, seeds of indigenous vegetables, if good and locally adapted stocks are available, may be produced by the growers themselves. The seeds of tomato, chilli, brinjal cucurbits, green leafy vegetables, peas, beans and okra can be produced conveniently and cheaply on the farm.

Vegetative propagation It makes possible production of trees of uniformly high quality. It perpetuates and multiplies desirable trees. The use of such trees makes fruit growing much less of a gamble.

Viability Ability to live, grow and develop.

Vibration of screen *See* speed of screen shake.

Vibrator separator Small model of vibrating separators have given excellent results on some difficult separation where the seed shows such minor differences in shape and surface texture that they could not be separated on other machines.

Vinegar Blemished fruits, wind falls or even cores and peels of fruit, which are normally discarded, can be used. Fruits containing 10 to 20 per cent sugar like grapes, apples, oranges, mangoes and dates, or cane juice, are utilised. The fruits are cashed, crushed if necessary, boiled, and the juice pressed out in a basket press. The sugar in the juice is added to 12 to 25 per cent either by diluting with water or by adding sugar. The juice is heated to almost boiling point, filled into a previously sterilised glass carboy, or a narrow mouthed glazed earthenware pitcher, to three-fourths of its volume. A small cake of pure wind yeast (obtainable from any winery or a food laboratory) is added to ferment the juice, the month of the vessel plugged with cotton wool to allow the gases to escape, and temperature maintained at 77 to 80°F, with frequent stirring. After the alcoholic fermentation is complete, when the juice shows about 1° Brix, it is stored for one to two weeks and allowed to settle. The clear liquid is siphoned into a clean carboy or pitcher till it is three-fourths full of mother vinegar (unpasteurised vinegar) added to it in the proportion of 1 : 10. A cork with two holes is fixed in the mount of the vessel to allow aeration and the temperature maintained at 77 to 80°F. A film of vinegar bacteria is formed on the surface and this should not be disturbed. The fermentation is complete in about eight to ten weeks after which the vinegar is siphoned off and preferably allowed to age for about six months. The clear liquid is then drawn off, filtered through several folds of cloth, heated in a open vessel to about 150°F, cooled and bottled. For its manufacture, culled fruits can be used and it is also of importance in the preparation of pickles, tomato sauce and other fruit and vegetable products. When vinegar is made from juice, the first step is alcoholic fermentation which is brought about by yeasts. Later the alcohol is acted upon by certain bacteria and turned into acetic acid. Vinegar can also be made from starchy vegetables, such as potatoes, the starch first hydrolysed into sugar. Vinegar does not ordinarily kill the organisms but prevent their development, hence its use in fruit preservation is of much value. It is commercial acetic acid. Bring this with *Vitis vinifera* viticulture, cultivation of grapes for the sale of grapes as fruit, and juice, as sultanas, currents or rasins or after fermentation as wine.

Viticulture Cultivation of grape vine for sale of its fruits, as a juice or rasins or after fermentation as wine. Its scientific name is *Vitis vinefera.*

Vitis vinifera grapes, it is susceptible to many fungal diseases.

Von thunen modem A model of agricultural and use that was devised in 1826 by Heinrich von Thunen, a prussialan landowner. It was designed to explain the factors governing the prices of farm products and the principles by which such price variations are translated into pattern of land use. Von Thumen envisaged a flat featureless plain (isotopic surface) with no variations in soil, fertility, relief or climate. This area is an isolated state in which one central city is the only source of demand and supply with no external trading.

Voralp The lower pastures of an alpine that they are reached before the mainalpine pastures on the higher ground.

Wacker process An industrial processing for making ethernal and other carboyl compounds.

Wadi A normally dry valley in a hot desert or semi-desert. It typically has steep walls and a flat bottom. Wadis sometimes have flash floods after heavy desert thundersome.

Wage records The form used for this purpose is :

				Regular workers' monthly or annual wages							
Name	Age	Date of employment	Date of termination of service	in cash	in kind	value	No.	Value	Clothing	Shoes	Total

			Casual employees			Meals' payment		
Name	Date	Operation Performed	in cash	in kind	Value	No.	Value	Total

Walnut juglans rigia Fam. Juglandaceae.

Warehousing Provision of a guaranteed storage services in return for a fee.

Warehousing operations The main function of these corporations is to acquire or build godowns or warehouses for the storage of agricultural produce.

Water absorption by humus Humus is very little soluble in water, but it can absorb gases as well as water. Absorption of water is accompanied by increase of volume. Lime has a precipitating action on humus. Humus also exhibits acid and base exchange properties and can thus form salts (which are electrolytes) and take part in cation exchange reactions. As a result, it reduces the possibility of damage from acids and alkalies.

Water course The water from sources is carried to the fields by flow due to gravity and is, therefore, cheap. It is diverted to the land to be irrigated through a system of canals, branch canals, distributaries and minors. From the distributaries and minors, water is given to farmers or a group of farmers through small structures known as soulets. From the soulets, water carried in small channels called water-courses to the fields. On a typical water distribution system, the quantity of water that finally reaches the field is about 55 per cent of the water that enters the canals. The loss which is mainly due to absorption, percolation, seepage and evaporation is of the order of 15 per cent in main canals, 10 per cent in distributaries and 20 per cent in water courses.

Water diving and water finder machines Sometimes help in locating a suitable site, certain people are said to be affected in a particular manner when they happen to walk on ground which has a good underground supply of water;. Some hold a forked twig or stick loosely in both hands, while walking along the ground to be explored. The water finder machine relies on the deflection of a magnetic needle suspended in it, which is moved by earth's current said to be produced by the action of the flowing underground water. Both these methods are said to have given successful results especially when worked by a person having a good knowledge of the surrounding country and its geological formulation.

Water duty Another term used in expressing the use of irrigation water is 'duty'. Duty denotes the number of acres covered by 1 cusec of water flowing continuously throughout the growing season of a crop. Duty, therefore, varies with crop, nature of soil, season and also a certain extent with the method of irrigation. If the water requirement of a crop is known, its duty can be calculated by dividing the quantity of water (expressed in pounds) which 1 cusec flowing continuously for the crop-season will supply by its water requirement (expressed in pounds).

Water logging Collection of rain water or irrigation water in low lying areas where it stays for a longer time making it impossible to use lands to be cultivated as due to accumulation of silt in the bottom of tanks or formation of hard-pan in the soil profile. The water is not able to move in any direction. Only a small part of the stagnant water evaporates due to surface-heat. Such areas are good for providing crops to Trapa.

Water lost in irrigation The loss represents waste of irrigation water. If the ground water table is already high, the excess water raises it still further and brings about water logging in the root zone. The plant roots are asphyxiated for want of air in the soil pores, and the plants wither and die. If water-logging continues over a long period, the dissolved salts are brought to the surface where, on accumulation, they turn the soil saline and some times alkaline.

Water meadow A low lying grassy area in humid-latitude regions usually besides a river or stream, that is permanently moist because of high water table and/or periods flooding. The associated soil type-meadow soil-produces lush grassland. This is grazed or cropped for hay. Some water meadows are deliberately irrigated to promote early growth.

Water relations Water forms 80–90 per cent of fresh weight of plants and animals both. It exists in plants as (*a*) solvent or true solution in vacuoles; (*b*) dispersal system and colloidal system, and (*c*) water held by pores of the cells. The factors which affect the distribution of water being these phases differ. The subject also include the mechanism by which penetration of water and nutrient elements into the cells is achieved and how the outward movement of cellular constituents is denied by the system.

Water requirement of crops Crops differ greatly in their water requirements. A large crop needs more used to the quantity of dry matter produced is called 'transpiration ratio.' This ratio varies from 200 to 500 in humid regions to twice that in arid regions. Rice, sugarcane, banana and potatoes have a high transpiration ratio. Wheat, barley, linseed and cotton have a medium ratio. Sorghum, bulrush millet and many of the small millets have a low ratio, which indicates their suitability for dry atmosphere, high temperature and strong winds increase transpiration ratio. Use of kind-breaks tends to reduce water loss.

Watershed The edge of a drainage basin. The watershed is commonly taken as a line joining the highest point on a boundary of a basin. Watershed below the ground surface do not always coincide with the watershed and water can leave from the drainage basin into another especially when there are steeply dipping sedimenting strata.

Wedge grafting Also called cleft grafting. It is used when the stock is a section of root. The lower end of the scion is cut to form a wedge with one side slightly longer than the other. An incision is made on the side of the rootstock at an angle of about 20°, deep enough so that the wedge of the scion may be inserted. Trying may not be necessary but waxing is.

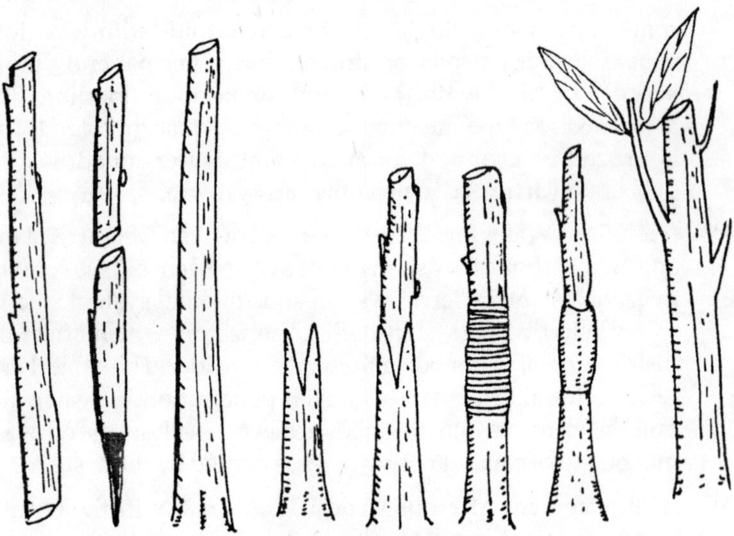

Wedge Grafting

Weeds All such plants which are not desirable to grow along with crop plants for their pernicious nature for growing good crops.

Weeding The operation related to the pulling out of the undesirable plants from the field.

Wet bulb temperature The lowest temperature to which an air sample can be cooled at constant pressure by evaporating water into it (the temperature of the air sample when it is saturated). It is measured on a wet bulb thermometer, the bulb of which is wrapped in muslin that is constantly kept moist.

Wet day A period of 24 hours starting at 09.00 hours GMT during which at least 1 mm of rain falls.

Wet point site A site at or near a source of water supply such as a stream, lake or spring where a settlement was located.

Wet spell A period of 15 consecutive wet days.

Wettable powder as herbicides It is a type of formulation in which a herbicide is absorbed generally on an inert carrier, together with an added surface-acting agent, a finally ground so that it will form a suspension when agitated with water. Simazine, atrazine, monuron, diuron and neburon are wettable powders.

Wheat (*Triticum* spp.) A cereal grass yielding a grain that can be ground into flour or meal. Widely cultivated and used in various forms all over the world.

A stuff of life for millions all over the world. In India, its whole-meal flour is used for making *chapaties, parathas, puries, halwa, uppamav* and a limited quantity for preparing loaf-bread, biscuits and confectionery. Chaff is fed to cattle. If occupies big area under cereals and a large average is under winter cereals. *Triticum aestivum* L., the bread wheat, forms the bulk of the winter crop. Macaroni wheat, *Triticum durum* Deaf, is the next important type; it is widely grown, *Triticum dicoccum* Schiible is grown to a limited extent.

Wheel burrows Common wheel burrows are very useful for gardens and orchards. The wheels may be mild steel or of rubber. The price of the wheel burrow is Rs. 90 to 100.

Wheel hoes These consist of a body, handle and a wheel. On its body, different attachments, such as shovels, a small plough, a small ridger and other types of blades can be fixed. The price of the wheel hoe is Rs. 30 to 40 with a weeding attachment. Other attachments can be purchased at extra cost. Being provided with a wheel, its is easier to work and is very useful for inter-cultivation in vegetable gardens and orchards.

Whip grafting Also called tongue grafting or spike grafting. The stock is cut at about 1 ft from ground and is cut diagonally about 1½ inch long, and same is made of the scion. In both the case, the size of the rootstock and scion is the same. The two are tight fitted, waxed, wrapped and tied at the union. Sometimes after making the diagonal cuts, both stock and scion are slit carefully half an inch from the end to a distance of about one half an inch. The two are then forced together so that the tongue enters into the slits. The graft is then waxed with or without tying. Both are used in root grafting.

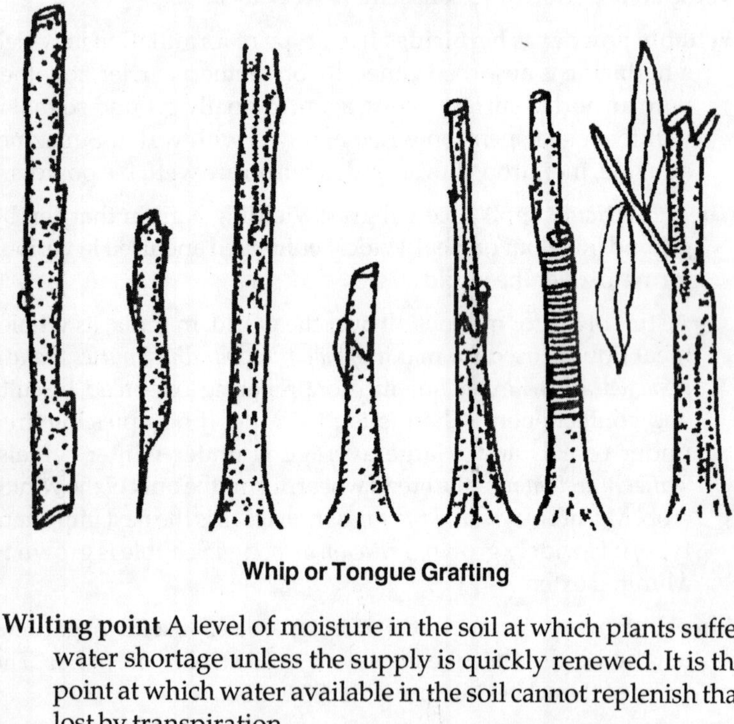

Whip or Tongue Grafting

Wilting point A level of moisture in the soil at which plants suffer water shortage unless the supply is quickly renewed. It is the point at which water available in the soil cannot replenish that lost by transpiration.

Wind The movement of air over earth's surface. It is almost totally horizontal, although there is a very small vertical component. The speed and direction are important in meteorology and are measured at all weather stations.

Windbreak As strong wind harm the fruit crop immensely by causing the flowers and fruits fall immaturely and also injure the young shoots. The one popular is raising a row of tall trees on the side of wind direction. It is a screen, natural or artificial, that is erected to check the force of the wind and give shelter to buildings or crops on the lee-side, it may be a line of shrubs or trees.

Wind breakers or shelter hedges Babusa of sorts, *Citharexylon subserratum, Parkinsonia aculeata, Prosopis juliflora, Salix tetrasperma, Sebania aegeptica* and *Tamarix articulata* are the hedges of this category.

Wind erosion The wind is responsible for three types of soil movement in the process of wind erosion. They are known as (1) saltation, (2) suspension, and (3) surface creep.

Windmills It is the cheapest source of power. As it depends upon wind, its working is intermittent and some kinds of storage of water may be necessary. The power developed is generally small (about 1/3 to 3.4 H.P.) wind velocity of not less than 6 m.p.h. is necessary to run a windmill. A 12 feet windmill mounted on a 70 feet tower raises on an average about 30,000 gallons of water per day.

Wind strip cropping It consists of planting tall growing crops such as jowar, bajra or maize and low growing crops in alternatively arranged straight and long but relatively narrow, parallel strips laid right across the direction of prevailing wind regardless of contour.

Wind velocity and soil erosion Similar reduction of velocity has been observed horizontally on the leeward side of shelter and wind brakes. When the wind velocity is 15 miles per hour, it is reduced to 60 per cent at a distance five times and height of the windbreak, and regains the original velocity at a distance 30 times the height of the windbreak.

Windward In the direction of the flow of wind.

Wind vane (Weather vane) An instrument that indicates wind direction.

Winnowers The winnowers are the machines in which a series of three to four sieves are also provided so that preliminary grading can be done. A fan is enclosed in a box like arrangement and is rotated fast. The sieves are also provided with a shaking arrangement. The steel winnowers cost Rs. 600 whereas the wooden winnowers cost Rs. 250 to 300.

Winnowing fans and winnowers To separate the chaff from the grain, the threshed material is put before a good breeze and the grain is allowed to drop by the weight of gravity and the bhusa is separated from the grain by few feet. In many areas, however, the natural breeze, at the time of winnowing is not available and the farmers have to wait for a number of hours and even days to get this breeze. Usually in these areas, there is a danger of rain and hail-storm during this period. The farmers are, therefore, very anxious to get the year's produce in their hours as early as possible. Some crude methods of producing artificial breeze have been used by the farmers, but recently winnowing fans similar to the ceiling fans have been introduced. These fans may be operated by hand or pedal, the number of revolutions are increased by using a pair of gears or sprocket and chain or belt drive. In some cases like the Nilokheri winnowing fan, no such drive is necessary. The size of the fan is such that it can produce good breeze. The object of using these fans is to produce only breeze and, therefore, no sieves are provided.

Winter annuals These are *Ageratum, Agrotis, Hollyhock, Sweet Alyssum, Anemone, Antirrhinum, Calendula, Companula, Ibris, Candytuft, Carnation, Sweet Sultan, Crysanthamus, Corn flower, Cineria, Clarkia, Coreopsis, Cosmos, Daisy, Dahlia, Dianthus, Gamolepsis, Sunflower, Helichrysum, Larkspur, Linaria, Linum, Lupines, Magnonette, Marigold, Nasturtium, Oxalis, Pansy, Petunia, Phlox, Garden poppy, Salvia, Sweet William, Saponaria, Stock, Sweet peas, Verbina* and *Wall flower*.

Winterbourne A stream that flow only during winter period.

Wood apple *Feronica limonia*, Fam. Rutaceae. A good root stock for sweet orange, brings dwarfness in the scion. Also known as Kaith.

Wood ash It is important potash supplying manure. It contains 3–7 per cent potash. It is used as fertilizer and is a source of potassium. It is mainly K_2O.

Working capital The part of a farm's capital that is used for carrying on of a business and is not invested in buildings or equipments. The amount of working capital required depends on the type of business. For an arable farmer, the working capital will consist of seed, fertilizer, insecticide, etc.

X

Xanthium strumarium Same control measure as for *Carthmus oxyacantha*. A weed.

Y

Yield Total production from an unit area in a crop season and from animals of milk in a lactation period.

Z

Zinc It is present in the plant tissue in about 5 to 100 ppm. It is also used as a catalyst in various chemical reactions in the plant tissue. Plants take it in the form of Zn^{++}.

It is a minor element used as a fertilizer. Its ionic form is Zn^{++}. Deficiency of this element is shown by the presence of yellow stripes on the leaves between the veins. Leaves become thickened showing a mottling condition.

A typical symptom of zinc deficiency in most crops is interveinal chlorosis of the foliage, particularly of the lower leaves. In severe cases, young leaves may also be chlorotic and lower leaves may turn brown to purple and die. There is shortening of internodes and stunted growth of plants. The buds of severely zinc-starved maize plants may become almost white. In trees, symptoms of zinc deficiency are known as rosettice, mottle leaf, little leaf and yellow. In citrus, the deficiency is usually known as mottle leaf and the affected leaves exhibit interveinal chlorosis, the vein remaining green.

Zingiber officinale Ginger. Its disease of rhizome-rot or soft-rot caused heavy damage to ginger crop.

Zinnia Some species suffer from the attach of black mould.